Kent Hufford

Reframing Technology

Kent Hufford

Reframing Technology

An Intellectual History of Thinking Beyond the Machine

DE GRUYTER
OLDENBOURG

Die vorliegende Studie wurde im Jahr 2020 als Dissertation an der Ludwig-Maximilians-Universität München angenommen.

ISBN 978-3-11-139611-8
e-ISBN (PDF) 978-3-11-139647-7
e-ISBN (EPUB) 978-3-11-139670-5

Library of Congress Control Number: 2024933896

Bibliographic information published by the Deutsche Nationalbibliothek
The Deutsche Nationalbibliothek lists this publication in the Deutsche Nationalbibliografie; detailed bibliographic data are available on the internet at http://dnb.dnb.de.

Cover image: A Cultivation of Technology (2024) Cover illustration by Cheryl Goods
Typesetting: Integra Software Services Pvt. Ltd.
Printing and binding: CPI books GmbH, Leck

www.degruyter.com

FSC www.fsc.org MIX Papier | Fördert gute Waldnutzung FSC® C083411

Acknowledgments

I would like to thank the following people and institutions, who in many ways made this book possible: the team at De Gruyter for their professionalism and cordiality, Susanne Dubeau, the Assistant Head at the Clara Thomas Archives & Special Collections at York University in Toronto for helping track down and find the “Mumford Manuscript,” the late Eric McLuhan and Michael McLuhan for their openness to my research and for allowing me to use the unpublished manuscript for my book, Professor W. Terrance Gordon for reading an early version of the McLuhan chapter and providing detailed and thoughtful suggestions and a warm scholarly embrace of my research project, the International Network for the Theory of History for providing the academic space to develop my research, Professor Hayden White for planting the seed for a metahistorical work, the staff at the Library of Congress Manuscript Reading Room in Washington, D.C. for access to the Stuart Chase Papers, Professor David E. Nye for accepting to be a second reader of this book and suggesting changes to the final chapter, Professor Miles Orvell for offering insightful thoughts on drafts of the manuscript, and the members of the colloquium at the Amerika-Institut in Munich for their feedback and comradery. Special thanks are in order for Prof. Dr. Klaus Benesch for being my “doctoral father,” for his patience, as well as for his guidance and encouragement throughout the project. Without his willingness to take me on as a doctoral student, this book would never have been written. Thank you to my cousins, aunts, uncles and other family and friends, to Tom Weaver for teaching me how to dream, to my dear grandparents William and Marjorie Hufford and Charles and Evelyn Tanner, to aunt Sandy for her enthusiasm and interest, to Bill and Joanne for their love and support, Loredana and Renato for allowing me to hibernate in their basement for a week to write parts of the McLuhan chapter, eat great food, and for tolerating my very broken Italian, to my brother Seth, my sister Dorie, my nephews Nevins and Miles for their hospitality, love, laughs and good times, to Dr. Emmett Goods for the unbreakable bond that we share, to my parents William and Dorothy, who through their constant love, support, and encouragement helped me grow and achieve with a steady, but gentle hand. And finally, thank you to my wife Giorgia for pretty much everything, which includes reading draft versions of all parts of the text and making wise suggestions, for her love and encouragement throughout the writing process, for her wonderful meals, and most of all for sharing time with me.

https://doi.org/10.1515/9783111396477-202

Foreword: Technology Is Not a Tool

This book was born out of an interest in history and philosophy. I was fortunate in that, while working on its outlines, a professor in Belgium named Berber Bevernage formed *The International Network for the Theory of History*. Participating in the network's conferences allowed me, at the time a part-time scholar working in the Dean's Office of the Faculty (School) for Languages and Literatures at the University of Munich, the opportunity to exchange thoughts with similarly interested people. It helped me feel less isolated in my thoughts that conventional ideas about history and who could write it, were changing. Reading and talking about texts and ideas in this and various other scholarly settings convinced me to develop my ideas a bit further and attempt a contribution of my own.

On first consideration, the book's major themes – history and technology – do not seem to have much in common. History is about the past and technology has a future orientation. But thinking beyond these clichés, history is also very much about the future in the sense that it orients the way we think about someone or something in the present, and technology, while concerned with the new, also has a conspicuous past. My personal bridge from thinking about history to thinking about technology was crossed when the acting Dean of the Faculty in Munich invited me to join his colloquium in the Department for North American Studies. In retrospect, for someone born in Pittsburgh, Pennsylvania, where century-old technologies and landscapes (oil refineries, coal patches, steel mills) dominate the region's history, uniting these two seemingly divergent subject matters in a book made sense.

Who might be interested in a history of writings about technology? Like many former industrial regions, Western Pennsylvanians had to reimagine and recover when technologies changed or technological methods became antiquated. Maybe reorientation has taken a bit longer in my birthplace because of the extent of the disillusionment: the dominant voices of the late nineteenth and early-to-mid-twentieth century were so confident in their telling of the story of technology. They assured progress and prosperity for all and espoused a faith in technology as a means to gain capabilities, increase power, or emancipate themselves and others from drudgery. Some went so far as to fuse technology with history, like in the phrases The Machine Age and The Great Technology, to signify a new and more promising era in human history. In fact, technology was, and is, a major source of identity in the steel town of Pittsburgh.

Recently, enthusiasm for a present and future bound to technology has been rightly met with skepticism. That is because, from today's perspective, the environmental and labor scars left behind by many so-called captains of industry are undeniable. This is true in the Rust Belt, but also for many other similar regions

https://doi.org/10.1515/9783111396477-203

across the globe. The stories we tell about technology have changed and people are right to question the role of technology in our own time. Are we gaining the greatest benefits from industry and technology? Is our situation any different from the past? For as much technology as we seem to have, why are we still working five days a week? My feeling while growing up in Pittsburgh, but also while writing this book, was that modern technology and quality of life did not necessarily align.

When thinking about technology during and after this book, readers are advised to remind themselves (rather often) that technology, as a concept, is made up. In other words, technology itself is a historical concept that was "invented" as a label after the fact to describe real-world changes. Yes, it is an essential component to modern life, but our ideas about it and what we think of it mask the fact that people have been deeply involved at every level – crafting, inventing, designing, manufacturing, working on, operating, maintaining, and using it. In other words, technology, in its broadest meaning, is not a thing. It is not a tool, as many people get wrong, but a culture of practice in the sense of work and labor. It is modified and can be modified at any time by intervention and alignment with human values.

Kent Hufford
Summer 2023

Contents

Introduction: When History Changes Technology

The Mechanistic theory of explanation turns upon the search for the causal laws that determine the outcomes of processes discovered in the historical field.
Hayden White, *Metahistory*

Did Technology Change History?

Living through the information age confirms what many people have long believed about modern technology: that its roots can be traced back to the industrial revolution, that it is a practical tool to achieve goals both big and small, and that it has brought about fundamental changes to society. To ask the question *did technology change history?* thus represents a starting point to examine assumptions about a common sense that has been constructed over the past one-hundred years.

The meaning of technology, you would think, is well-established.[1] But the reader might be surprised to realize how relatively difficult it is to actually say what technology is or involves. The word itself can be easily traced, but the concept behind the term has a complex history of changing meanings, all of which continue to shape the way in which it is currently used and understood. Like most large terms, technology means different things to different people and these meanings can be at odds with one another. In everyday language, technology is often reduced in meaning to only signify the newest product, like a computer or other digital device, which is at the end of a development cycle. Yet limiting the concept to this understanding misses so much about technology. For example, when a technology is being designed, like an AI application, software as a service, or a smartphone, technical knowledge and skill are used to create it. Specialized technical knowledge is obviously built on prior generations of understanding and must be tediously maintained and advanced in theory and in practice. Skill in application likewise depends on the working environment in which a product is developed and modified. For these reasons alone, it should be obvious that the meaning of technology involves much more than just the brilliant application of scientific knowledge. Any understanding of the concept must necessarily include people, their cultures, and their labor.

1 See appendix 1 for a short rendering of the rather recent development of technology as a subject of historical inquiry in North American discourses.

https://doi.org/10.1515/9783111396477-001

Zooming out to gain a broader view, technology can also be understood as a *set of practices to create and use material and non-material things* (Schatzberg, *Technology* 2). This distinctive way of seeing technology as activity[2] makes even more sense, for instance, once the design phase ends. A technology's development and historical context are subject to change at all times. In fact, a story about technology becomes all the more interesting once a finished product enters use. Who used it and to which ends? How was it modified over time? Did it privilege or negatively impact a certain group of people? These are just a few of the many questions that can be asked once the concept of technology is put in various historical contexts.

Similar to technology, the term history seems to have an accepted meaning. However, attempting to nail down exactly which history is implied quickly reveals the term's complexity. History can have at least four basic meanings: (1) *historical reality*, meaning all that happens at any given moment without being recorded, (2) a *discipline*, meaning the subject matter of historians, a specific and particular (3) *historical account* of people, events, ideas, things, systems, etc., and what we understand in plain terms as (4) *the past*, the empty space in which we can formulate and situate a meaningful story.[3] If we think of history as historical reality, the first example above, the opening question becomes *did technology change our reality?*, which is a very pressing question indeed. Typical of conventional accounts of technology is a kind of hagiography of inventors or discoveries that changed the way we did something. As the story goes, improving technology in and of itself is a worthwhile venture. The technology in these formulations is often a product or application of useful thinking, and it is here that one can detect further assumptions about how we tell stories about technology. For one, pursuing technological advances is a largely beneficial endeavor for society. It is developed to improve something and historical reality bears out such improvements. Secondly, it is something external, a tool or system outside our body (or something that can be implanted into it) which we can control. Or can we?

The effect of these stories about technology is to reinforce the idea that things have indeed changed our reality. So, what is wrong with this idea? A labor or social historian might stress the significance of people or movements that brought about change. But even in these accounts, technologies are frequently seen as external tools to be used or that have been abused, as opposed to a process or order of human activities. It is therefore useful to reflect on the question *did technology change history?* and its possible answers to help clarify one's relation to forces

2 Technology as activity can easily be understood by listing activities like crafting, inventing, designing, manufacturing, working, operating, maintaining, and using (Mitcham 210).

3 For a more in-depth explanation of these four different uses of the term history, see appendix 2.

and systems that appear greater than people and their ability to change them. Why is this important? Because the future is ominous – nuclear war, global pandemics, environmental collapse – and because each new generation is forced to learn how to govern technological and ecological transformation that seems novel, uncontrollable, and barreling toward the unknown. Maintaining humanity within this context is both challenging and overwhelming.

To address such feelings of vulnerability and futility, this book argues that asking questions like *did technology change history?* is misguided and leads instead to a framework of understanding that disempowers a majority of people to accept either the status quo, or even worse, debases potential sources of empowerment. This may seem vague, but the argument against this simple question is that trying to answer it diverts our attention from basic, systemic solutions to human needs and forecloses possibilities and futures we might find more attractive. How is that? With danger on the horizon, common sense tells us to plan ahead and develop new collective technologies. The logic goes that if technology is a tool that must be invented, we need smart people and more and better technology to avoid war, sickness, and environmental destruction. If, however, technology is actually something different, like a social process, would it not be smart to shift from a set path of innovation to allow for new possibilities? This may be what we need to open up the future from the one that is implied by a certain fixed view of what technology is. What is at stake is not only a more egalitarian conception of technology, but also the ability to counterbalance the reality effect of dominant, rigid narratives that would suggest otherwise. Common sense is not static. It changes constantly and is corrupted and enriched by various sciences and philosophies. This book argues that we need to change our common sense about how technology changed or did not change history.

To this end, the following pages and chapters offer alternate understandings of the technological past and present that can provide for greater autonomy from living conditions often not on balance with basic human needs. The (hi)stories in this book are representative of a shift in frameworks about how the story of technology has normally been told, a revision to the common sense. It happened over the course of the twentieth century and increasingly rejected the notion that technology has determined our past, controls our present, or dictates our future. They attempt to combat rigid narratives and old frameworks, which are sticky, and continue to exist into a projected future.

Technological Determinism and History's Reality Effect

In accounts where technology is thought to strictly control, shape, or even dictate human behavior, it is said to be a case of hard determinism, and has been roundly dismissed by cultural and literary scholars as an explanation of historical change.[4] Yet in spite of such harsh criticism, the idea has yet to expire.[5] So what causes it to endure, despite being outdated, disproven and at worst, being considered a false and reductionist explanation of technology's effects in society? Put simply, why do people continue to believe that technology can determine historical outcomes?

The modern genesis of the idea is often ascribed to Karl Marx, who, to use an illustration from one of his early books, wrote that "by changing their mode of production, men change their social relations" (K. Marx 48–49). Taken literally, this statement was used to demonstrate that Marx supposed a hand-mill to be the direct cause of feudal society and a steam-mill the direct cause of industrial, capitalist society. So not only has he been associated with the idea of determinism, but it has been attributed to him in its most extreme form (Sowell 71). Yet as Donald MacKenzie explains in his essay on Marx and the machine, while deterministic sentiment is certainly to be found in his work, characterizing him as a determinist "in any strong sense" is "difficult to sustain" (MacKenzie 193). In reality, the role of technology in capitalist society first proposed by Marx is "more subtle and compelling" (Noble xix). Marx has been mischaracterized. He believed in historical change as a result of a dialectic between modes of production and society rather than one causing change in the other in a simple, linear way.

More than a century and a half after Marx, the debate rages on. Now, these kinds of accounts that contain a direct relationship between technology and society are referenced by the term *technological determinism.*[6] In fact, it is one of the central questions that have concerned technology scholars over the last half century (Fouché xxi). Although there are different interpretations of technological determinism and some ambiguity as to its referent, it generally indicates the significance of technology in causing or determining historical outcomes. But rather than precisely

4 For an in-depth forum on the subject see Smith and Marx's volume *Does Technology Drive History?*

5 See the discussion in Sally Wyatt's "Technological Determinism is Dead; Long Live Technological Determinism" where Wyatt acknowledges both the longevity (staying power) of the idea and various forms that it has taken.

6 For a general introduction to the subject matter, see the four-volume series on *Technology Studies* published by Sage (2008), which contains original versions of seminal papers in the field that conceptualize technology (volume 1), theorize technological change (volume 2), examine the political ramifications of leaving technology unchecked (volume 3) and analyze non-dominant views of technology (volume 4).

defining technological determinism, it is important to acknowledge its significance as an argument. Whether it comes in different forms[7] or has various faces,[8] technological determinism has staying-power as an explanatory narrative of historical change. As a way to justify certain outcomes, technological determinism's simplicity and linear nature have undoubtedly aided its acceptance and longevity, as well as its aptitude at explaining modern human experiences. This effect is compounded by increasing and ever more present modes of history and historical consciousness. The idea is close to everyday reality and has the ability to infuse meaning and agency into historical phenomena and provide common sense explanations. This further serves to increase its legitimacy, which may explain why scholarly insights need the help of larger frameworks of understanding to carry meaning into greater public discourse.

Such a dearth in scholarly relevance may be due in part to what Martin Davies has called "the redundancy of history in a historicized world" ("The Redundancy of History in a Historicized World" 335). On this view, what results from a certain kind of history is the "aura of a finished world." This line of thinking also suggests that sophisticated historical explanation lacks the resonance of enthusiastic phrases like "innovating toward the future" or "technology drives history." Simple narratives contribute to a finished world, or what Herman Paul labels the "completed past" (Paul and Runia 21–23). This is a past that is stable and seemingly cannot be changed. It is also a past that precludes, occludes, or renders in-

7 In arguing that scholars should take technological determinism seriously, Wyatt distinguishes four types of technological determinism: "justificatory, descriptive, methodological and normative" (167, 174–176). The first type, justificatory technological determinism, is used by actors to justify and strengthen their argument/positions. The second or descriptive form is a "mode of explanation" that has been denounced by Misa (1988) and Smith and Marx (1994) as an inadequate explanation of historical change. The third type is practiced by STS scholars and is a "tool for helping us understand the place of technology in history." The final type of technological determinism, which the author attributes to Bimber's Norm-Based Account, holds that technology has become so complex that it can no longer be socially controlled. In sum, Wyatt rejects technological determinism as an "adequate framework for understanding the relationship" between society and technology but reminds readers of its relevance as a justification for action by "decision makers."

8 Attempting to prove that Marxist history is not technologically determinist, Bimber describes three "faces" or accounts of technological determinism which include the "Norm-Based Account," the "Logical Sequence Account" and the "Unintended Consequences Account" (218–220, 226). The first sees the "replacement of ethical norms with goals of efficiency and productivity" in that social control is lost over technology. He assigns Lewis Mumford's writings on technology to this account. The second account amounts to hard determinism in that it claims that technology "exercises causal influence on social practice." This second version is what Bimber associates with true technological determinism. The third, final face of technological determinism is a weakened form, which observes that technology is only partially "responsible for determining certain social outcomes."

significant many alternate understandings. Historical explanations of change fueled by simple narratives make the proliferation of technology into all parts of life digestible as a historical phenomenon. Compounding an overload of history is that fact-checking or more complex historical context often fail to move people towards a particular moral sentiment.

As a dominant cognitive norm, history, in its simplest forms, produces a historicized world that induces "cognitive and behavioral ambivalence, a paralyzing oscillation between comprehension and apprehension, between aspiration and misgiving" (Davies, "The Redundancy of History in a Historicized World" 336). Given the degree of saturation in a historicized world, anyone attempting to explain the increased presence and supposed influences of technology in everyday life over the last two centuries is wise to address discourses that surround technological determinism in its justificatory form.

The actual stories of the past need to be distinguished from the specific form in which they are expressed (Munslow, *Narrative and History* 9–10). This is an important distinction because conflating the past's content and history as a particular form of telling the past without explicitly addressing an author's narration leads to the so-called "reality effect" of history. The problem with this effect is that it can "project the world as a natural outcome of rational necessity" (Davies, "The Redundancy of History in a Historicized World" 335). In the case of technological determinism, an idea emerged to fill the semantic void that was created by societal and cultural changes starting in the nineteenth century (L. Marx, "Technology" 562–563). As an explanation for historical change, technological determinism can be and has been used as a justification by actors that prefer certain outcomes.[9] In other words, the view that technology drives history can be exploited to rationalize systems of power. As the "dominant mode of intelligibility," history as a form of telling the past can have the effect to "automatically normalize" such a historical justification (Davies, "Cognitive Inadequacy" 1–2).

A concrete, historical example of this phenomenon can be found in the official guidebook to the 1933 Chicago World's Fair. In the text of the guidebook, "going back a century" meant depicting how rough and crude life was in 1833 and contrasting it with a vastly better and more "felicitous" life in 1933. This account of the past framed so-called achievements to have been the result of "a century of progress," and conflated Chicago's growth as a city with the growth of science and industry. The theme perfectly served its purpose: to promote consumerism of industry's products and to cement the idea that humans needed to adapt to technology and not the other way around. This latter point can be summed up in a heading

9 Similar sentiment can be found in Wyatt (2008).

from the guide, "Science Finds - Industry Applies - Man Conforms" (Official Guidebook of the Fair 11–15).

Given the reality and normalizing effects of history, how can dominant and hegemonic narratives like technological determinism be mitigated? The next section provides some background into a shift in frameworks that put technological change into a new historical context. This endeavor happened after many decades of worker-organized reckoning with industrialization, mechanization, and system building. The struggle from below, to recognize the efforts of labor activists and social reformers in the late nineteenth and early twentieth centuries, was to enfranchise groups in the decision making of industrial plans and projects. If the driving force behind changes to society was not people, but some external force called technology, its specter was useful to the business of a select few, not the many. By the late 1920s, stock-taking and evaluation of society's ills called for a new type of historian of the technological.[10]

A Shift in Frameworks

This book is about a shift in frameworks. It attempts to specify a change to the historiography of technology and embed that change in historical context. Since the onset of what is commonly referred to as the Second Industrial Revolution (1870–1930), scholars have debated the impact of machines in society. As machines became more complex and part of large technological systems, there was a growing sense amongst social scientists that machines had altered the trajectory of human history. In the early twentieth century, in order to make sense of material changes, some intellectuals still used a mechanistic explanation according to which technology drives history. This view imbued technology with causal agency as a directing and shaping force of historical change. In contrast, other scholars attempted to understand the effects of technology within a larger framework in which they contextualized change by overcoming the separation between technology and sociocultural factors. More recent research in the field of the history of technology has demonstrated that the underlying concerns of mechanistic explanations are deterministic and can lead to essentialist concepts of technology. Scholarly consensus has consolidated instead around a more contextualist way of

---

10 Other intellectuals at this time were reinterpreting the past for their own emancipation, not only from the effects of the industrial system, but from the effects of slavery and subsequent forms of oppression, like W.E.B. DuBois had in *Black Reconstruction*, published in 1935.

interpreting change that attempts to explore the ways in which humans and technologies interact or shape social and cultural history.

While technological determinism has been academically debunked, what still requires investigation are the circumstances in which the move away from deterministic narratives and toward contextualist interpretations of technological change took place. This development corresponded to a shift in historiography from frameworks of necessity to frameworks of contingency. In other words, scholarship transitioned from linear and causal understandings of technological change to more complex interpretations dependent on a constellation of social and historical factors that needed to be contextualized. This book locates the genesis of this historiographical shift in the context of North American interpretations of technological change at a specific time – roughly the first thirty years of the twentieth century – during which scholars were just starting to come to terms with the effects of industrialization in society. Focusing the analysis on this period is crucial because it is then that technology, a concept now central to explaining modernity, acquired its present meaning.[11]

In order to trace this shift, the current work argues for the need to extend the historiographical analysis beyond traditional and purely historical accounts. It proposes instead to identify a corpus of non-professional historians whose perspectives expand the understanding of this crucial turn. This analysis starts with Thorstein Veblen, a philosopher, economist, and social theorist, and proceeds to investigate the writings of economists Harold Innis and Stuart Chase, and of literary and cultural critic Lewis Mumford. Together, these scholars' theoretical contributions helped advance the shift from necessity to contingency through accounts of technology that diverge from deterministic understandings. Coming from different academic backgrounds and each adopting particular relations to technology and to the technological past, these scholars are representative of a trend in historiography that argues for the importance of context and contingency to interpret technological change. What is significant about the selected authors and their works is that they challenged accepted explanations and narratives of cultural change in and around technology, then an emerging and contested concept.

The current study then continues to include an analysis of Marshall McLuhan's work, which belongs to a slightly later phase of the historiographic shift toward frameworks of contingency. By investigating McLuhan's aim in writing a historical analysis of the end of print culture and a guide to understanding media, it will be

11 As Erik Schatzberg has noted, "when historians now address 'attitudes toward technology' before 1930, they are employing an analyst's category not used by the historical actors themselves" ("Technik Comes to America" 486).

demonstrated that the Canadian theorist sought to raise cultural awareness about technological environments in order to help his readers better understand and adjust to a recent change in dominant forms of communication. Similarly, by locating and describing his notion of causality, it should become clear that McLuhan's work attempts to interpret a constellation of factors that together create an environment from which technological change emerges. The characterization of McLuhan as a non-determinist proposed here is an endeavor not only to correct typical misinterpretations of his work, but also to show how they arose and how they can be avoided. Part of the evidence provided as justification for this conclusion was taken from an unpublished and previously once-referenced manuscript in which McLuhan confronts Mumford's supposed misunderstanding of his writings on technology.[12] The untitled manuscript, here provisionally referred to as "The Mumford Manuscript," is a 17-page reaction to Mumford's monograph *The Pentagon of Power* that can also be read as a justification of McLuhan's own approach to critique technology.

Chapter Overview

This book opens with a first chapter that provides the reader with a general background about the historiographical challenges that historians in the second half of the nineteenth century faced when trying to describe material change. The chapter then traces the theoretical origins of technological determinism and the narrative according to which technology drives history and is followed by a section that describes a crucial shift toward contingency in concepts and frameworks to explain historical change. The discussion turns to Thorstein Veblen's critical role as a mediator between German discourses and a new American concept of technology. In a North American context, this concept was used both to critique the industrial system and capitalism, but also to signify social progress. The authors at the center

12 In conducting research on McLuhan, I happened upon reference to this manuscript in Terrence Gordon's *McLuhan: A Guide for the Perplexed* (2010), which can be found on page 135. Due to the vast amount of literature on McLuhan, I cannot be absolutely sure that other scholars have not referenced or analyzed this manuscript. Eric McLuhan, Marshall McLuhan's son and collaborator on many academic projects, had not been aware of its existence before personal correspondence with the author, writing "I am quite gratified to have a copy of this wee essay" (Eric McLuhan, email message to the author, 16 April 2018). Additionally, McLuhan expert Terrance Gordon wrote in a personal email that, "I am delighted, of course, that the soon-to-be-famous file 287 has been unearthed" (Terrance Gordon, personal email to the author, 16 April 2018). Based on these responses, there is a distinct possibility that the manuscript had simply been misplaced in a box of rather mundane business correspondence.

of the current study are briefly introduced and situated within the context of two key historical phases, the first, which involved a response to industrial mechanization in the first half of the past century, and the second, which immediately followed World War II. By reflecting on the context that these authors responded to in their work on technology, the current analysis contextualizes their reactions in relation to a change in discourses.

The second chapter builds on an argument made by scholar Eric Schatzberg in the recent monograph *Technology – Critical History of a Concept* (2018). Schatzberg claims that Thorstein Veblen appropriated the discourses around the German word *Technik* into American-speaking scholarship. He posits Veblen's legacy as a dichotomy divided into cultural and instrumental understandings of technology. This book expands on that line of analysis by arguing that Veblen provided a template for later scholars to interpret cultural understandings of technology. In effect, this template could be either used to show how human values shape technologies or abused to view technology as an instrumental means to an end. The chapter ends with an assessment of Veblen's work by Canadian economist Harold Innis, who summarized Veblen's thought in the form of a detailed literature review.

The third chapter examines the writings of U.S.-American economist Stuart Chase. In the chapter, Chase is shown to have followed in Veblen's theoretical footsteps concerning technology and its relation to society more than others have acknowledged. Initial emphasis in the chapter is placed on Chase's background leading up to and after his ouster from a government position in the Federal Trade Commission. Focus is then directed toward his early analysis of "waste" in the industrial economy as an engineering problem. Chase's general concern is the improvement of the society. With regard to this goal, technology is viewed as both a help and a hindrance. The discussion in the chapter advances to examine another of Chase's books that attempted to survey various aspects of modern life in a technological environment. Chase takes an engineering approach to not only look at the effects of technology in society, but also to assess their advantages and disadvantages. For Chase, the industrial situation is fragile, set on false premises, and should be changed to allow for more collective wealth. At the end of the chapter, Chase's contributions to early discourses on technology are considered in the context of his impact on Lewis Mumford.

Although inspiration for much of the analysis in the chapter was taken from two of Chase's early books, in comparison to the other authors in this study, supplementary information on Chase is rather scarce. For this reason, additional sources and evidence for the arguments made in the chapter were taken from the Stuart Chase Papers, a special collection in the Manuscript Division of the Library of Congress in Washington, D.C. This assortment of original documents, newspaper clippings, magazine articles, pamphlets, booklets, notecards, and personal correspondence greatly aided the author's ability to provide the reader with a more reliable understanding of Chase.

Chapter four explores Lewis Mumford's ideas about technology and technological change within a cultural context. Mumford referred to his subject matter as a "complex," which suggests his tendency toward a contingent understanding of technology.[13] Of particular note is that when Mumford began writing about technology in the late 1920s and early 1930s,[14] he oscillated between using the term technology and another less familiar form that can be derived from Greek, *tekhne*,[15] or its Anglicized version technics. In his much later two-part history *The Myth of the Machine* (1967/1970), Mumford plots out social, aesthetic, and moral relations to technics. By making technics a historical subject, his aim is to reveal its negative goals, which he sees as the exercise of power and decision-making behind what he terms the "megamachine." In doing so, Mumford holds out hope that mankind can save itself from destruction by its own hand. The current analysis begins by exploring Mumford's early years, his influences, and some of his formative ideas. Through a sampling from his first few books, the chapter details Mumford's embrace of a framework of contingency and illustrates an early version of his core historiographical methodology. This detail helps establish Mumford as a main theoretical node in the book's overarching attempt to trace a historiographical shift away from frameworks of necessity. The next two sections of the chapter provide some context leading up to the writing and eventual publication of Mumford's 1934 book *Technics and Civilization*. This monograph, one of the first scholarly attempts to explicitly historicize understandings of technology in English, serves as the subject of a close reading. The analysis argues for two different readings intended to enrich the existing scholarship on Mumford with regard to historical theory. The chapter ends with a final examination of how Mumford thought technological change occurred and with what purview historians should study it.

---

13 According to Webster's dictionary, the word "complex" can be understood to mean "a whole made up of complicated or interrelated parts." Today one might include the notion of network in Mumford's definition.

14 In a chapter on art in *Wither Mankind* (1928), Mumford mentions *technology* seven times and *technics* only once. The naming of his monograph *Technics and Civilization* (1934), which he researched for during his European trip in 1932, reflects the conceptual influence German discourses of *Technik* had on his subsequent word choice.

15 Much later, Mumford justifies a return to the "classic Greek term *tekhne*" because it "characteristically makes no distinction between industrial production and 'fine' or symbolic art" (*The Myth of the Machine* 9). On this understanding, technics is "related to the whole nature of man" furnishing a healthy relationship between "objective conditions" and "subjective needs." Mumford understands the term 'technics' to be a "broadly life-centered ecological complex."

The fifth chapter examines the work of Canadian scholar Marshall McLuhan, who radically redefined technology, conflating it with both conventional and unconventional forms of media. His aim in treating media/technology as a historical subject was to make humans aware of the effects that they have on our sensorium. Only by studying media/technology, so he thought, can one combat their physical effects and save oneself from their numbing effects. The analysis begins by detailing Mumford's criticism of McLuhan's ideas in order to demonstrate that the two men were reacting to two different historical phases of North American discourses on technology. Despite this change in discourses, McLuhan's approach to technology is positioned alongside Mumford, as part of the shift in historiography. McLuhan, primarily as a scholar of literature, also worked within a framework of contingency in his inquiries into perception and communication. Two of McLuhan's influences are highlighted. McLuhan's training in poetics is established under the tutelage of I.A. Richards and is followed by a short analysis of Harold Innis's research framework to investigate the relationship between technology and society. McLuhan's decision to adopt this framework of technology's past is characterized as significant. The chapter continues by detailing McLuhan's often misunderstood methodology and sources and is followed by an analysis of several of his first articles on technology. The discussion turns to McLuhan's first major book on technology's past, *The Gutenberg Galaxy*, and attempts to demonstrate that McLuhan was arguing against deterministic ways of thinking. A section follows that details McLuhan's move from theory to praxis in which he wrote a guide to the effects of media – his famous book *Understanding Media*. Significantly, McLuhan is shown to posit the new media environment as a challenge to culture. The next section of the chapter describes the concept of causality that McLuhan emphasized in his work with regard to technological change and is followed by his appeal to historians to drop their practice of espousing value judgements and become more like perceptive artists that displace conventional ways of seeing the world. The chapter continues with the presentation of an unpublished manuscript in which McLuhan criticizes Mumford for his approach to technology and his lack of a theory of communication and change. The last sections of the chapter highlight McLuhan's approach to technology and summarize the argument that rejects misinterpretations of his work as overly deterministic.

The final chapter connects the authors in the current study with the emergence of formal and institutional inquiry into narratives of technology. Subsequent scholars have fully embraced frameworks of contingency, and I detail several specific ways in which frameworks of necessity, narratives of progress, and technological determinism have been rejected at the end of the twentieth century. The last sections of the book underscore the importance and relevance of its analysis. The authors main contributions are revisited and distilled into a brief synthesis.

A Note on the Approach Taken in this Book

Beyond historical and cultural studies affinities, the current inquiry assumes a literary/metahistorical perspective.[16] The approach employed in this book is literary because the selected authors did not strictly write in the historical, meaning realist mode. It is also metahistorical because it was primarily inspired by the work of Hayden White. Working metahistorically in this case means assessing the work of the selected authors through the reoccurring theme of technology in their texts, but also engaging with the philosophy of history by working "in between" disciplines. Drawing on classic examples, philosophers of history have traditionally had two objects of study: historical reality (*res gestae*) and historical thought (*historia rerum gestarum*). The first type has a long tradition and eventually became associated with Hegel in the nineteenth century. This type of reflection has been called the material, substantive and/or speculative philosophy of history. The second type of philosophy mainly came about in the nineteenth century and is referred to as the formal, analytical, and critical philosophy of history (Paul and Runia 7). Such reflection mainly looks for laws, patterns, and regularities in historical thought.

Recently, the old bifurcation that characterized the field of philosophy of history has changed, along with a broader interest in history researched, displayed, and performed outside the academy. Scholar Herman Paul notes two tendencies that emerged from the inaugural conference (2013) of the International Network for Theory of History: an expansion of the type of history that historical theorists are interested in, meaning both inside and outside the academy (historical discourses, experiences, and memories) and the integration of previous work in the field ("Relations to the Past" 451). In his introduction to the field, Paul argues for a united term to reflect these newest developments: introducing the term "historical theory" (Paul and Runia 1–16). In renaming the field of research, Paul tentatively claims, the "broadly shared set of beliefs" that compose the current research agenda that historical theorists should pursue is summed up in Mark Day's phrase "relations to the

---

16 It has been suggested that the older field "Philosophy of History" adopt the name "Historical Theory" to combine the research agendas of various traditions within field. Although this supposedly new field has become more interdisciplinary and includes more objects of analysis than its predecessor, Historical Theory does not "signal a break with the nineteenth- and twentieth-century authors" like Ranke, Comte, Spengler, Toynbee, Berlin and Popper. In essence, the renaming of the field attempts to form a more inclusive research agenda through the recognition that people "relate to the past not only with their brains (their reason), but also with their heart (their emotions and life-orienting values) and their hands (their encounters with material remains from the past and their attempts to make the past serve political ends)" (Paul and Runia 12–15).

past." The most salient point of this new vocabulary for studying history is that the various ways of relating to the past (epistemological, political, moral, aesthetic, etc.) interact when a historian ventures to understand, interpret, or experience the past. Paul is quick to point out that these relations are teleologically oriented, being "defined by the goals (*teloi*) they seek to acquire" ("Relations to the Past" 453). These goals "overlap, interact and influence each other." Table 1, taken from Paul's book, shows a sampling of the various relations to the past and their corresponding aims.

Table 1: Relations to the past with corresponding aims.

Relation	Aim
Epistemic	Knowledge, Understanding
Moral	Goodness, Justice
Political	Exercise of power, Decision-making
Aesthetic	Beauty, Coherence
Material	

© Herman Paul. *Key Issues in Historical Theory* (2015).

Emphasizing a scholar's relation to the past helps uncover the framework which he or she uses to understand the past. This is why, when analyzing the relationship between technology and history vis-à-vis the selected author-historians,[17] the current book uses the vocabulary and research agenda proposed by Paul, which extends to non-professional historians, to explicate "what the past does to people and what people try to find in the past" (Paul and Runia 31).

Implicit in the field of historical theory is its nominal contrast to history. The former involves abstraction, theory building and recognition of alternative narratives based on the omissions and facts established by the latter. The intent of this book is not to investigate how technology affects historical realities, but to analyze historical thoughts on technology that contest a framework of necessity commonly referred to as technological determinism. Using the vocabulary of the relations to the past, this book strives to demonstrate how each of the author-historians analyzed here elaborates an alternative understanding of the past and responds to an inquiry into technology's role as a cultural phenomenon in history. Ultimately, this study allows for the formation of a group of authors that raise and attempt to respond to questions about human and cultural aspects of technology that contest the notion of

17 Described "as the 'I' in history" Alun Munslow argues that authorship is fundamental to history as a form of re-presenting the past (*The Future of History* 148). In doing so, he further underscores the notion that history is "*primarily* an authoring activity."

inherent progress in technological change. Taken collectively, their inquiries into the technological past attempt to bring choice, balance, and cultural and environmental awareness back to human life upset by rapid and seemingly autonomous change.

This final point indicates that reframing the past to demonstrate the contingency of technology is of vital interest to us all. When history changes technology to reveal the true complexity of the subject, it should become increasingly evident that we could use some basic orientation for our current unenviable situation of war, sickness, and climate disaster. Automation, performance, and power must be devalued to make room for more life-sustaining uses and practices. In abstract terms, this means the solution to sustaining life on earth is to understand technology in terms of culture, and actively govern the uses to which technologies are put by society. Merely balancing positive and negative effects would allow humans to gain renewed confidence in our ability to create an enriching environment for ourselves and others. But beyond that, histories of technology need to be linked up to a more practical past.[18] With this more concrete goal in mind, I invite readers to familiarize themselves with the understandings of technology and the technological past that are presented on the following pages. It is my hope that readers are able to connect the main points of this book to fundamental questions and issues of technology and culture in the present, whenever that may be.

18 The usage of the term is taken from Hayden White's description of philosopher Michael Oakeshott's distinction between the "historical" and the "practical past" (*The Practical Past* 9). By practical past, White refers to notions about the past that help people to solve practical problems in their everyday life.

Chapter 1
Honeycombing the Past

Finding Meaning in a Rapidly Industrializing World

History had no use for multiplicity; it needed unity; it could study only motion, direction, attraction, relation. Everything must be made to move together [. . .].
Henry Adams, *The Education of Henry Adams*

By the eighteenth century, a subtle transposition of values had begun to take place, as technics itself began to occupy a larger place. If the goal of technics was to improve the condition of man, the goal of man was to become ever more narrowly confined to the improvement of technology. Mechanical progress and human progress came to be regarded as one; and both were theoretically limitless.
Lewis Mumford, *The Pentagon of Power*

It is important that we understand the Nineteenth Century, for the patterns of behavior and of thinking which it produced honeycomb our actions and our thinking today.
Stuart Chase, *A New Deal*

Modern Historiography and the Material Changes of the Nineteenth Century

In his famous autobiography, American historian Henry Adams questioned the antiquatedness of his nineteenth-century education for the twentieth century, a world that he understood to be vastly different from the one he had been born into. Keeping in mind that the book is narrated in the third person and is widely understood to be written in an ironic, self-deprecating tone, the thought Adams expressed reveals history's fundamental problem, which is keeping up with the pace of change. As the first epigraph above suggests, Adams retrospectively recognized that a change in historical education and practice would be necessary to describe and narrate the complexity that resulted from massive increases in power during his lifetime. As an older gentleman, he must have felt he needed different concepts and methods to narrate history if it were to remain an academic and scientific discipline. The simple stories, the political history, the linearity and straightforward narrative had to be retired.

In Europe, before the academic discipline of history was supported within a national framework, history writing, or historiography, was mainly documented

Note on chapter title: As a verb and in this context, honeycombing involves penetrating the spaces of the past and filling them in with meaning.

https://doi.org/10.1515/9783111396477-002

through institutions linked to the church. Yet, by the end of the early modern period and certainly by the nineteenth century, these institutions had been swallowed up by major empires. In a provocative observation, Henry Adams assessed the damage: the "stupendous failure of Christianity [has] tortured history" (Adams, *The Education* 472). "If," he asked, "after nineteen hundred years, the world [is] bloodier than [before]," what is the purpose? If meaning was not to come from the church, where would it come from? What would be narrated as meaningful and who would be charged with telling collective stories?

Since at least the publication of Giambattista Vico's *The New Science* in 1725, scholars and philosophers had posed similar questions about the meaning of history and encouraged a rethinking of methodology and direction that is characteristic of the development toward modern historiography. The theory of salvation and practice of theodicy would eventually be challenged by less universal forms of thinking about the past. By 1850, nationalism had replaced Christianity as the main unifying political force in Europe. More emphasis was placed on individual orientation during a finite period of time, which included categories of collectives like class and nation. A new orientation and method emerged as historians were now employees of a state, and no longer in the services of a church.

When professional historians first wrote about material changes that characterized the mid-nineteenth century, they could employ the mechanistic mode of explanation[19] to achieve a specific *telos*, or goal, in writing history which was to educate their readers about a political past filled with affirmation of national growth and development. Framing past reality in this way would help justify why great personalities and their choices led to industrial and material innovation, which in turn propelled history toward a seemingly imminent and positive outcome.

The research phase of the so-called historical method was refined during this era of historiography, referred to within the discipline as the era of historicism.[20] Researching was understood to involve discoveries in archives, objective judgement of source materials and deduction from general principles to specific examples. The writing phase required the mind of a great storyteller to place the causal links along the appropriate historical timeline. As an academic discipline, the professionalization of historiography coincided with the idea that, in distancing itself from the arts,

---

19 In his groundbreaking 1973 monograph *Metahistory: The Historical Imagination in Nineteenth-Century Europe*, Hayden White characterizes mechanism as a reductive world hypothesis. As for a definition of the mechanistic mode of explanation, White explained that "The Mechanistic theory of explanation turns upon the search for the causal laws that determine the outcomes of processes discovered in the historical field" (16–17).

20 Although the term historicism can be employed differently, here it refers solely to an era of historiography from around 1850–1960. For clarification see Kolmer 44 or Jordan 40.

it "is or should be a science" (Lorenz 393). By the turn of the twentieth century, scientific historiography was the dominant, institutional form to think and write about the past.[21] If historical questions could be addressed to a professional and satisfactory standard and then presented for criticism to other university-trained historians, it was thought, something close to an objective truth could be ascertained.

To quickly get a sense of historiography at the beginning of this era, one can imagine a historian's answers to three basic questions:

- How did material changes affect society (civilization)?
- What caused those changes?
- To what end were those changes enacted?

Consider the first question (how). A common perception amongst historians in the nineteenth century was that society had changed mechanically, sequentially, linearly, and purposefully. Their likely response to the second question (what), was that human choice or free will could cause those changes. And finally, an answer to the third question (to what end), summed up in the word progress, provided historians with a desired effect. Their task was to research and write about the circumstances, actors, and actions that directly caused it.

This method of doing history would suffice until material changes as part of what is referred to as the Second Industrial Revolution (1870–1930) became so drastic, particularly in the United States, that historians would be challenged to describe historical events. The problem of how to describe the past and of how to make sense of these changes would require new concepts and new discourses. Terms like art and science had become too narrow in meaning to adequately describe industrial modernity (Schatzberg, *Technology* 72). And historical agency, often reserved for chosen inventors and innovators, was increasingly ceded to a new driving force.

How – A New Dynamic

By the late nineteenth century there was a greater understanding of the complex in the natural sciences, and a basic recognition of what could not be perceived by the human senses. Increasingly, sophisticated tools were needed to measure how

21 The term "Scientific Historiography," as defined by Chris Lorenz, concentrates on three agreed "constitutive ideas" of this type of historiography, namely "truth," "objectivity" and the "critical method" (393–403).

things happened. In contrast, theorizing of the relationship between cause and effect experienced a bit of lag within the confines of the historical guild. Most historians continued to write monographs with the same narrative toolkit as if unconcerned about the fundamental breakthroughs in knowledge occurring in other fields. In an ironic take on his own discipline, Henry Adams expressed uncertainty in the accepted understanding of historical causes and effects:

> Historians undertake to arrange sequences – assuming in silence a relation of cause and effect. These assumptions, hidden in the depths of dusty libraries, have been astounding, but commonly unconscious and childlike; so much so, that if any captious critic were to drag them to light, historians would probably reply, with one voice, that they had never supposed themselves required to know what they were talking about (*The Education* 382).

Adams reasoned that the perplexities of historical change could benefit from understandings of cause and effect in the natural sciences. Like other historians of his day, he was interested in the dynamics of change and he hoped that more rigorous scientific investigation would result in a law or set of laws that would be more adept at explaining and predicting human history in pursuit of ever-increasing power. It is not that he distrusted historians to explain historical sequences and outcomes, he merely thought that employing a scientific method would further the professionalization of the discipline.

What – A New Force

In place of a purely sequential understanding of historical causality,[22] Adams offered a "dynamic theory of history" according to which progress was defined as the "development and economy of Forces" (*The Education* 474). Man, in his search to assimilate power, is a force; nature is another, and together with a multiplicity of other forces they are constantly "reacting upon each other" to determine a sum (478). Adams must have assumed, even if sarcastically, that one could understand history as a "problem of mechanics." Much like a mathematical function, inputs (forces) could be placed into a formula in which a set of outputs (sum of forces) would result. If these values could be ascertained through historical research, one could understand how and why a force developed in the past.

With the advent of the scientific method and the introduction of inventions like the compass, humans became capable of widening "the range of contact between nature and thought" to create a "new variety of mind" (*The Education* 482). Increas-

22 Causality is understood here to mean a certain type of explanation reflecting an understanding of the relation between cause and effect.

ingly, keen observers and experimenters had become aware of a "new class of supersensual forces" (486).[23] Adams confidently pronounced that man had been able to keep pace with these forces, even though such discoveries, figuratively speaking, had changed the equation to a degree that Adams felt compelled to add a law of acceleration to his theory of history. The law asserted that present and future forces would increase exponentially in speed and complexity:

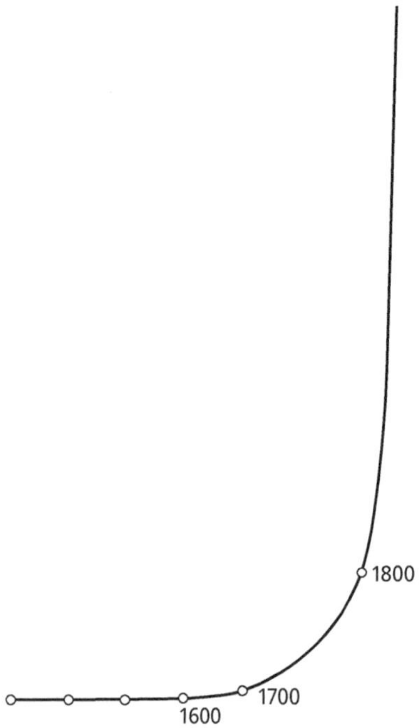

Figure 1: The curve of history, by Henry Adams.

As the "curve of history" (Figure 1) clearly illustrates,[24] human choice as a factor in history was in danger of being completely overwhelmed by the acceleration of these forces. Adams thus warned that society would require a "jump" to a "new so-

23 Discussing such forces, Adams lists that "science has proved that forces sensible and occult, physical and metaphysical, simple and complex, surround, traverse, vibrate, rotate, repel, attract, without stop; that man's senses are conscious of few, and only in a partial degree; but that, from the beginning of organic existence his consciousness has been induced, expanded, trained in the lines of his sensitiveness; and that the rise of his faculties from a lower power to a higher, or from a narrower to a wider field, may be due to the function of assimilating and storing outside force or forces" (487).
24 The chart is taken from Adams's 1919 book *The Degradation of the Democratic Dogma* (292).

cial mind" in order to react to them. "Thought" he maintained, "must enter a new phase subject to new laws" (*The Education* 498).

In correlation with the growth of industry into the early twentieth century, as machines, industrial processes and technical skill and knowledge increased in specificity and prevalence, a new interpretation of history began to appear in the social sciences in which human choice was less salient and greater emphasis was placed on how a new force itself could power historical change. This jump in agency from human to technology as a prime motor of history was a profound conceptual change. It conspicuously provided an alternative rendering of who or what controlled developments in history.

Why – A New Type of Explanation

Alongside recognition of technology as a new force in history, a new causality, or type of explanation of historical change, also emerged between the two world wars. As Eric Schatzberg has pointed out, the modern concept of technology was largely absent in nineteenth-century economic theory (*Technology* 157). For example, Karl Marx did use the word *Technologie* in his most famous work, *Das Kapital*, but he employed the term to propose a history of technology, or a historical study of the material basis of societal organization and the production process. Marx used other words, like machinery, to try to nominalize economic changes within the context of the rise of modern industry, but there was a general lack of scholarly attention paid to the role of technical skill in production. This void in theory and the concurrent semantic confusion left scholars in the early twentieth century, like progressive historian Charles Beard and sociologists Talcott Parsons and William Ogburn, to introduce a reductive concept of technology in their academic theories of explanations of society. In effect, these men relegated the concept of technology to a type of instrumental explanation for change. Beard left a focus on economics as "the principle motor of history" to switch to technology (Schatzberg, *Technology* 139). Parsons conceptually limited technology to a type of instrumental rationality to achieve a specific goal, while Ogburn equated its meaning with invention (167).

A specific example of this line of thinking can initially be found in Ogburn's *Social Change with Respect to Culture and Original Nature* (1923). In a section on "social maladjustments," Ogburn asks whether the ultimate "accumulative nature of material culture" has led to an "increase in inventions," which in turn may be the proximate cause for the "rapidity of social change" (200). He later explicitly argues for material culture as a "source of modern social changes," reasoning that the speed of material changes alone is enough evidence to give it meaningful status as a historical agent: "The very fact that material culture is to-day undergoing such rapid changes means that it is significant as a cause of social phenomena" (277). Ten years later, in a booklet entitled *Living With Machines*, Ogburn unabashedly disseminates the idea that machines create economic and social problems. In a section called "machines create problems" he is ultimately led by his reductive concept of change to ask what the state should do with "business organizations that the machines create?" (8). In another section called "what do machines do to us," he interprets the work of Stuart Chase, who at the time was writing about social affects in the machine era, to imply that machines have "greatly modified our social life" (11).

Such interpretations of the conceptual meanings of technology made it possible to respond to the question "why" or "to what ends" with verbs like to mechanize or to systemize. Improvements, regressions, and other changes to what Ogburn referred to as material culture, or what now might be called technology, were strongly factored into changes to the human condition. As a result, human progress was wedded to technological progress. They had merged into the same narrative.

An effect of the merging of human and technological narratives was its instrumental use by capitalists and engineers to change how America worked. This process of building, mechanizing, and systemizing is best exemplified in so-called improvements in factory production, respectively named after Henry Ford and Frederick W. Taylor. Ford became wildly rich and known for mass production, while Taylor pioneered scientific management. By disciplining production, both forms of organization were to an attempt to streamline and control every aspect of an industrial plant and its workers. The status quo was to be replaced by efficiency, meaning no movement or process was wasted in these function-dominated environments (Hughes, "Machines, Megamachines, and Systems" 110). Using this logic, one could justify a concern with systemic efficiency, as opposed to worker's rights and needs, as most beneficial to technological progress, and so the narrative goes, to the greater good. Tracing the origins of this peculiar logic is the focus of the next section.

A Historicizing of the Direct Causal Link between Technology and Progress – The Rise of a Concern

The role of technology as a driving force in American life is at the core of the narrative of progress that has dominated the American historical imagination.[25] As David

25 The direct causal link between technology and society was not always seen as progressive. Regressive relations to industrial changes can be observed for example in social reformer Florence Kelley's *Modern Industry in Relation to the Family, Health, Education, Morality* (1914) where, in a chapter on "Modern Industry and Morality" Kelley laments the lack of recognition of the causal relation between industrial business practices and poor working and living conditions. Kelley writes

Nye has observed, it was particularly during the Jacksonian era, that the rural United States underwent a major transformation with regard to industrialization, which then began to be understood as a particularly American phenomenon (*Technological Sublime* 45). On this understanding, social and national goals are achieved through the genius of scientific inventors, the building of sophisticated machines and the perfection of applied knowledge and technique. Among the first and lesser-known attempts recorded in American history to change society using what could be considered technological means was a large-scale endeavor that sought to provide "everything desirable for human life . . . in superabundance" (Etzler 1). This particularly fantastical promise depended on a machine that would free men from labor to create a communal paradise "within 10 years." Spelled out in great detail over two-hundred pages, German-born engineer John Adolphus Etzler advanced the mathematical calculations for a system of communal living and plans for a machine that could harness energy from natural sources like the wind, ocean tides, waves, and the sun. The idea that machines can bring about change may seem redundant to the modern reader, but for a man born in 1796, the "science of mechanics," then in its infancy, provided the first real means to utilize the awesome potential of machines to cause positive (and normally omitting or ignoring any negative) changes in the human environment, harness the power of nature, and even further human progress. Seeking an improvement to the human condition of his time, Etzler detailed the method to achieve such ends in his book titled *The Paradise Within Reach of All Men*, which was published in 1833 in Pittsburgh.[26]

In seeking such progress through a machine, Etzler is an explicatory figure because his promise can be used to demonstrate where historical theory can help interpret deeply rooted notions of technology and technological change. Once formidable and quasi-fatalistic, such linearity in the causality between technology

that "The causal relation of industry to the present evils in social life and personal character has been slow to compel recognition, slower in this country than in Europe" (134–38). The main difference to the scholars mentioned in the previous section was that Kelley did not attribute agency to technology and explicitly asked whether "industrial morality" included a tendency to "restore to the people who work a share in the ownership and control of the tools of industry" and promote "the enactment of the industrial code."

26 In essence, Etzler proposed what one might call a technical solution to societal problems in order to rid men of the necessity to labor for their needs. Trained as an engineer, possibly influenced by Hegelian ideas of progress, Etzler tried multiple times to garner support for a project to organize a society in the New World according to early socialist principles. He and his partner Conrad F. Stollmeyer founded the Tropical Emigration Society in 1844, which, when at its height, was the largest socialist emigration project of the early Victorian era with 1,558 members. The men and their Society ultimately failed to sustain a community, in part because of disputes surrounding land purchase (Kirchberger 94), but surely also because Etzler did not deliver on his technical promises.

and change is now viewed with due skepticism; the direct link from machines to progress has been severed. It has become irrational to think that machines alone can bring about the revolutionary changes to society that Etzler had once envisioned.[27] Indeed, much has been written to disentangle the notion of progress from its supposed mechanical enablers. Already in the mid-nineteenth century, as Leo Marx observed, the distinction between "mechanical means and political ends provoked ideological backlash" ("Technology" 566). An example of such adverse reaction is a sarcastic essay penned in 1843 by Henry David Thoreau, who took aim at Etzler's version of a mechanically achieved paradise and his "chief faults" of seeking "to secure the greatest degree of gross comfort and pleasure merely" (302). Acknowledging Etzler's predictable failure to fulfill his ten-year promise, Thoreau commented snidely that "Alas, ten years have already elapsed, and there are no signs of Eden yet" (300).

Complementary to Etzler's error in conflating machines and progress is the apparent lack of historical interest in his story.[28] Historians of technology can be excused for relegating Etzler to an insignificant figure as not much is known about his life. As for the chosen site of his paradise, Etzler drew upon the potential of America as a utopia. With its specifically Puritan origins, projections of America's future included the idea of history as a progressive drama under god's direction. Importantly, as Howard Segal makes clear, the cultural background of Puritanism is the ideological base from which critiques of American technology should be viewed (4). While his methods to achieve a type of "technical utopia" on American soil may have failed, Etzler's plan to use machines to improve human conditions is remarkable, not only because the idea still seems strikingly contemporary, but also due to the ideological complexities that it contains: machines enabling social progress, machines driving progressive change.

So where did Etzler draw the idea to take things into his own hands and attempt to forge a more egalitarian utopia with a machine? During his lifetime, ideas about history in what is now Germany were changing from Hegelian-inspired idealism to the beginnings of what later became known as the philosophy of historical materialism. More specifically, the absolute principles of the former were beginning to give way to the mechanistic construct of the latter in which real progress could be achieved through changes in the means of production and ownership.[29] This change in the conception of history meant that the religious/fatalistic notion of

27 Viewed as a technical utopian, the thrust of Etzler's own argument relies heavily on the salience of reason (every man's ability to think) and intelligence (knowledge of useful things), as well as the avoidance of the "stupid adherence to custom" (Etzler 89).

28 For a recent economic interpretation of Etzler's story see *The Great Delusion* by Steven Stoll.

29 For a comparison of these two different philosophies of history see Jordan 24–31.

god's will unfolding was replaced by the idea of progress caused by man and material. There was a sense that one could and should actively change history by unlocking knowledge and applying it to the world. Humans were free to enlist the help of machines and Etzler argued that men of reason could change society in the right direction. Trying to use machines to attain societal aims, Etzler may have drawn on a mutation in the understanding of historical agency from celestial to human origins. It is this general shift in historical consciousness that serves as the starting point for new narratives of progress of improving society through technology.

Some fifteen to twenty years after Etzler's attempts,[30] scholars describing changes in their nations made a switch from writing historical narratives inspired by the "romantic imperative" to those about "progress and materialism" (A. Smith, "Seven Narratives in North American History" 62–66). These descriptions were mainly in reaction to the "hard facts of material growth," assisted by new machines, and the "tangible results and achievements" that helped them understand, define, and differentiate themselves when compared to peoples of other nations (62–66). As such, the conceptual framework for the idea that machines drive history had now been established, but it was some years later that the idea became more fully realized by the realism[31] and historicism[32] of the first professional historians. As Hayden White summarizes, "In general, then, the 'realism' of nineteenth-century historical thought consists in its search for adequate grounds for belief in progress and optimism" (*Metahistory* 48). While nineteenth-century historical accounts sought evidence that the choices of certain "great personalities" led to superior outcomes, twentieth-century historians and sociologists went one step further. They linked ma-

30 The Tropical Emigration Society was founded in 1844 and by the 1860s, not professional historians, but "mainly self-taught observers" began to challenge the "romantic imperative" in historiography (A. Smith, "Seven Narratives in North American History" 64–65).

31 Hayden White explains realism as a philosophical position: "To be a 'realist' meant both to see things clearly, as they *really* were, and to draw appropriate conclusions from this clear apprehension of reality for the living of a possible life on its basis. As thus envisaged, claims to an essential 'realism' were at once epistemological and ethical . . . to express desire to *be* 'realistic,' then, must reflect a specific conception not so much of what the essence of 'realism' is as of what it means to be 'unrealistic' (*Metahistory* 46–47). The problematics of a realistic approach to reality are much the same as those contained in the notions of 'sanity' and 'health.' Such notions are more easily defined by what men of a given time and place recognize as their opposites, 'madness' and 'sickness.' So, too, the specific content of a given age's conception of 'realism' is more easily defined by what that age as a whole took to be 'unrealism' or 'utopianism.' "

32 Due to the "proliferation of different and conflicting uses of the term" Robert D'Amico defines the central meaning of historicism as a "family of arguments for thinking knowledge about the past is possible" (249–51). In this context, I use the term according to Stefan Jordan's definition as "an epoch in German historiography between 1800–1960 that is characterized by historians who viewed history as a progressive development driven by 'great men' " (40).

chines and emerging industries to the concept of progress to solidify the idea that an autonomous force called technology, largely divorced from a common person's imagination and control, represented a significant force in determining historical outcomes.

Viewed through a contemporary understanding of historical theory, the "desire to uncover an underlying driving force that determines the teleological course of history" is characteristic of modern historiography (Susen 137). On this view, historical developments are said to be obliged to "necessity"[33] so that any Enlightenment-inspired search for meaning and direction is dictated by modernity's "mechanistic conceptions of history" (139). Such underlying principles bolster the causal determinism attached to a force like machines or technology when scholars attempt to describe change. The implication is that a modern search for meaning and causality within this framework is accompanied by the rise of a concern over whether technology drives history.

Far from being a historical issue, discussions about technology's effects are currently hotly debated. Beyond Thoreau's early transcendentalist criticism of Etzler's goals, there has been an ongoing reckoning with the framework of necessity built into modern historiography since at least the early twentieth century (Susen 169). The ascent of realist, scientific historiography[34] and the coinciding end of utopia in history[35] did however present an opportunity for a few concerned scholars not only to reject further attempts to reconstruct certain concepts of technology as a fixed historical reality,[36] a theoretically tenuous position known as naive realism,[37] but to also take on a different approach when interpreting an emerging technological past.

33 According to Susen, the five central assumptions that historical change is "determined by necessity" are: 1. historical lawfulness 2. historical predictability 3. historical linearity 4. historical teleology 5. historical universality (137).
34 "Scientific Historiography," as defined by Chris Lorenz, was detailed above.
35 See Hayden White's "The Future of Utopia in History," in which he posits modern utopian thinking as history's other to problematize accepting realist arguments of reality (12–19).
36 A historical reality that includes the idea of technology as man's lawful, predictable, linear progression as a tool-making animal towards a universal goal (teleology).
37 Munslow believes that few (academic) historians still ascribe to this position. Conversely, the so-called "realist-reconstructionist" view of history is seen as the dominant and "culturally acceptable way" to "engage with the past and its knowable truth" (*Narrative and History* 12). It is so widespread among the "history-consuming general public" that he somewhat ironically names it the "common-sense" position.

A Change in Concepts, A Change in Frameworks

Today, serious scholars no longer explain historical change using the mechanistic mode or historical laws. Nor do they imbue technology with superpowers as an agent of change. The phrase technology drives history is recognized both as an outdated narrative of progress and as a mechanistic theory of explanation. Current approaches to history and technology studies tend to focus on the way in which particular aspects of technology and culture interact towards contingent, fleeting, and contextualized outcomes. In a paper theorizing technology and modernity, scholar Philip Brey concludes that:

> the major insights of technology studies have been that technologies are socially shaped and society is shaped by technology, or, alternatively, society and technology co-construct each other. They are not separate structures or forces but are deeply implicated in each other. Technological change is moreover not a linear process but proceeds by variation and selection, and technologies have interpretive flexibility, implying that their meanings and functions and even (according to social constructivists) their technological content are constantly open to renegotiation by users and others (54).

With regard to historical theory, one can say that a shift has occurred during modernity from what can be understood as a framework of the *necessity* of uniform change using grand narratives that assumes a linear path of progress, to a framework of the *contingency* of a possibility for change within small narratives, which may or may not have a fitting ending. Some of the first attempts to explain the social processes and conditions that are now considered characteristic of modernity – industrialization, rationalization, and emancipation[38] – represent a good example of the fractured and contradictory discourses at the core of this shift in historiography.

To write about and make sense of massive cultural change, early twentieth-century scholars had to grapple with terminology for their concepts. As alluded to previously, the term that eventually won out to help conceptualize these changes did not exist with its current broad meaning. In the nineteenth century, the word technology was primarily associated with education, referring to the study of techniques or to polytechnic universities. And the concept of technology seemed to be missing altogether from nineteenth-century economic theory with regard to the "role of technical knowledge, practices, and artifacts in production" (Schatzberg, *Technology* 157). To describe the role of technical knowledge, skills, and their application played in bringing about material changes, scholars used words

38 Susen understands these and other key features to be both a precondition for and result of modernity. For an explanation of these and other features see the section entitled "Key Dimensions of Modernity" (13–16).

like invention and machinery, but also a terminology that belonged to a once ill-reputed subset of a larger, more general conceptual domain. In the past, various modifiers had been attached to art to indicate changes in material culture, from the vulgar arts (ancient) to the mechanical (medieval), the useful (manufacturers' eighteenth-century contrast to the fine arts) or industrial arts (nineteenth century). This variation in terminology to refer to what is now collectively called technology can be observed in the naming of so-called technical museums in major European cities: the Musée des Arts et Métiers, which translates as Museum of Arts and Crafts, founded in Paris in 1794, the Museum of Manufactures, founded in 1852 in London, and the Deutsches Museum von Meisterwerken der Naturwissenschaft und Technik in Munich, founded in 1903.[39] The latter name contains the German word *Technik*, which was particularly influential to developments of the English concept of technology.[40]

In the late nineteenth century and into the early twentieth century, German scholars had vigorously debated the supposed intrusion and meaning of *Technik* with respect to German culture (*Kultur*).[41] American scholars were aware of these discussions and tried to translate and appropriate discourses around the term into their own books in English. In contrast to German discourses of *Technik* and *Kultur*, in a North American setting, the question of whether technology was or was not part of culture was redundant due to a conflation of the two through the narrative of industrial progress. There was one concept that could be used to describe industrial modernity in English, but there was a difficulty in using a modifier with the collective term art. The singular term industrial art had associations with the fine arts, while the plural, industrial arts, came to refer to "vocational training in manual trades" (Schatzberg, *Technology* 121). As Schatzberg has indicated, English-speaking scholars at the time "lacked a general concept of technology comparable to that of *Technik*" (122). Using various words or phrases like machine age, technique or state of the industrial arts, American scholars often struggled to translate or even mistranslated *Technik*. Moreover, they included or disregarded aspects of the concept so that tensions emerged amongst understandings of its place in modern society.

One writer in particular, Thorstein Veblen, did theorize technology in relation to culture in an innovative manner. As a number of scholars have written, Veblen played a vital role in adopting the German discourse on *Technik* and "making *tech-*

39 Information was taken from each museums' official website – 1. https://www.arts-et-metiers.net/musee/visitor-information 2. https://www.vam.ac.uk/collections/the-va-story 3. https://www.deutsches-museum.de/.

40 For knowledge of the influence of discourses surrounding *Technik* on *technology*, I am deeply indebted to Eric Schatzberg's conceptual history of technology.

41 For a good summary of the German discourse on *Technik* see chapter 7 of Eric Schatzberg's *Technology: Critical History of a Concept*.

nology a keyword of modern American culture" (Schatzberg, *Technology* 125).[42] Schatzberg explains how Veblen developed his idea of the term into a "sophisticated concept" that drew from both "cultural and instrumental aspects of *Technik*" (130). In contrast to the largely positive and progressive depictions of industry and mechanization associated with nation-building narratives, in his writings Veblen developed a critical framework to untangle technology from culture by differentiating what he would call the "instinct of workmanship" from the "absentee ownership" of industry. This distinction between the technological knowledge and processes and privately-owned business interests made it possible for future scholars to envision technology unhinged from dominant economic practices. It is worth exploring Veblen's critique,[43] not only because of the new concept it contains, but also for the historical moments which stimulated it and its later reception.

Technology – An Anti-capitalist Term or Embedded in Capitalism?

The genesis of Veblen's anti-capitalist concept occurred mainly in response to the Gilded Age (1870–1900). Starting through mass women's movements and organized labor, a discontented population sought to secure greater equality for most citizens.[44] Criticism of the economic and political order by university professors soon followed. Historian Charles Beard, described the overdue support from academics: "naïve acquisition and enjoyment were coming under the scrutiny of sophistication" because "social criticism had crept into scholarship" (Beard, Charles and Mary Beard 429–30). Veblen's academic work reacted to the prevailing economic system at the time, here broadly understood as American capitalism with a market economy. At the beginning of the twentieth century, Veblen published two major books that were critical of the modern industrial system, *The Theory of Business Enterprise* (1904) and *The Instinct of Workmanship and the State of the Industrial Arts* (1914). It is important to note that Veblen's essays "On the Nature of Capital," along with other critical writings, were composed during this time. Many of these essays were of academic nature and critiqued economic theory.

As the meaning of industrial technology took on heightened relevance while the first large-scale experience of mechanized and global warfare raged, Veblen collected his observations in the form of an academic comparison of industry in

42 Primarily Eric Schatzberg, Leo Marx and Ruth Odenziel.

43 A brief synopsis of his writings is provided in the next chapter.

44 Acrimoniously, most citizens often did not include African Americans or other non-white groups.

England and Germany. The book, *Imperial Germany and the Industrial Revolution*, was published in 1915, but Veblen had largely written it before the war had broken out. The destruction that resulted from the First World War evoked a fierce response from more prominent voices. For example, in a 1915 article written for the *New York Times*, English science-fiction writer H.G. Wells framed the conflict as a "breaking point" for civilization (Wells). Famous for his descriptions of technological futures, Wells promptly insisted that (Western) peoples of the world had a "choice" to limit the "development of scientific warfare" and attend to the "inherent" dangers that new inventions like the submarine and airplane latently possess. Wells argued further that weapons and techniques of war developed for World War I should be viewed as the latest evidence needed to shake "inattentive minds" into rethinking the goals of technology as a whole. In a section entitled "A Blind World," Wells commented that "never was [there] so blind a superstition as the belief that progress is inevitable." The author thus makes a strong argument against continuing to apply "human intelligence" to the "mechanism of war," a habit that will result in "progressive destruction." History, he suggested, could also provide "ample support" to prove his point.

After the war, economic theory turned to historical practice as Veblen's individual essays on the industrial system were collected and published in book form as *The Place of Science in Modern Civilization and Other Essays* in 1919. By the early 1920s, Veblen's final collection of essays, *Engineers and the Price System* (1921), and his final book on the state of the industrial system, *Absentee Ownership* (1923), had begun to disseminate. These first analyses of the unrestrained capitalist-industrial system were read by many, including Stuart Chase, who admired Veblen's work.[45] At the start to his own career as a writer, Chase penned a critique on the lack of management in industry in 1922, a booklet for the League for Industrial Democracy. He extended that idea to book length in *The Tragedy of Waste* (1925). In both writings and based on Veblen's theories, Chase criticized wastes in consumption, in labor resources, in the methods of production, as well as in natural resources. Not to be mistaken for overbearing scientific or business management, Chase's corrective to the status quo of industry was a populist attempt at critique, as will be shown in a later chapter.

Given the background described above, why would a concept that was born out of a critique of capitalism be turned around by later historians and social thinkers to signify societal progress? This question appears to identify a paradox, but one that is rather instructive. The application of a general word like technology or technological

45 Chase wrote the foreword to the 1934 Random House edition of Veblen's *The Theory of the Leisure Class*.

to describe profit-making enterprises is paradoxical because, as Veblen argued, the economic system in place was not fully beneficial to the common individual. Quite to the contrary, the conditions during which his critical notion of technology and its conceptual framing took shape (the Gilded Age and World War I in particular) implored a critical assessment of the industrial system and its effects in society. So, what can be learned in discerning this inconsistency? As Leo Marx has indicated, "it exemplifies the congruence of technology and corporate capitalism" ("Technology" 575). That is to say, the meanings of words and concepts, even ones newly created by Veblen, are still socially constructed. Leo Marx cites Raymond Williams's observations about "the circularity, or reflexivity, involved in the social construction of keywords – words like *culture, industry, democracy* – which came to serve as historical markers for the periods when they acquired new, fundamentally altered, meanings" (Williams qtd. in L. Marx, "Technology" 575). The same dynamic, Marx continues, can be observed with technology:

> it is now customary to single out the transformative power of technology as the defining characteristic of the era when, in response to remarkable advances in the mechanic arts, the meaning of the word technology underwent a radical change. The term that formerly had named a field of study now referred to the society's entire stock of technical knowledge and equipment (575).

Williams's analysis is supported, with regard to technology, when the boundaries are blurred between the "material (physical, or artifactual) components of these large socio-technological systems and the other, bureaucratic and ideological components," or when the boundaries are erased between the "whole technological system" and the "surrounding society and culture" (575).

Based on the above analysis, it seems fair to argue that technology can be understood to be both anti-capitalist and infused with transformative power through its entanglement with capitalism. Significantly, starting in the late 1920s and early 1930s, several of the first critical histories of technology began to be written by North American authors, some of whom, in particular Stuart Chase's *Men and Machines* (1929) and Lewis Mumford's *Technics and Civilization* (1934), can be directly linked to Veblen and his radical understanding of technology free from the "vested interests" in modern industrial society.[46]

46 According to Veblen in *The Vested Interests and the State of the Industrial Arts*, a vested interest is a "legitimate right to get something for nothing, usually a prescriptive right to an income which is secured by controlling the traffic at one point or another. The owners of such a prescriptive right are also spoken of as a vested interest. Such persons make up what are called the kept classes. But the kept classes also comprise many persons who are entitled to a free income on other grounds than their ownership and control of industry or the market, as, e.g., landlords and other persons

Historical Critique of Technology by "Prophets of Contingency"

The period of time in which technology emerged as a key word has also been recognized as the first phase of a delusion with necessary links, such as those between technological and economic systems.[47] In his "Epilogue on the History of the Post-Historical Moment," Simon Susen divides the period that led toward the emergence of postmodern thought in the past century into "three key historical phases," the first of which starts precisely in the 1930s. According to Susen, these phases occur respectively in (1) the early to mid-twentieth century, (2) the mid-to late twentieth century and (3) from 1989 onwards in response to historical events at the time.[48] In his assessment, social and historical events in the twentieth century "undermined the grip of certain Western beliefs including that reason, emancipation, science, truth, progress, [. . .] are necessarily linked and – more importantly – that the optimistic trust in the civilizational triumph of these ideals will manifest itself in the development of an even more just and empowering society" (Susen 169–170). At this point, it is fair to question the relevance of key historical phases in the emergence of postmodern thought to early histories of technology. The connection is in the authors analyzed in this book, whose works were published at the beginning of the first phase (Chase and the early Mumford) and into the second (the later Mumford and McLuhan).[49]

Writing in response to industrial mechanization since the nineteenth century, which he identified as a "radical transformation in the entire human environment," and to the "megatechnics" of the "dominant minority," meaning the rise of fascism and state socialism, American author Lewis Mumford attempted to understand the "dominating role played by technics in modern civilization" (*Technics and Human Development* 3; *Technics and Civilization* 4). Mumford's first book on the subject, *Technics and Civilization*, was both novel and significant. Instead of viewing technol-

---

classed as "gentry," the clergy, the Crown – where there is a Crown – and its agents, civil and military" (161–62).

47 The metaphor "Prophets of Contingency" is borrowed from Yemima Ben-Menahem. As she makes clear, the framework of contingency in history can be used to change perspectives not only to be a "lever for social critique and political change, but also a manifesto for greater human independence in every realm" (129).

48 This historical categorization can be found at the end of chapter 4 of Susen's book *The Postmodern Turn in the Social Sciences* (169).

49 Works published during the third key historical phase, while relevant, exceed the scope of the current book. Having introduced Stuart Chase's relevant works in the last section, the next two sections will mainly focus on representatives of the first (Mumford) and second (McLuhan) key historical phases for the purposes of this book.

ogy as an independent progressive force in history, Mumford's lasting contribution from this first foray in scholarship on technology is to have incorporated the subject into broader discourses of civilization. Mumford's important legacy to the history of technology was put into perspective by Donald L. Miller, who called Mumford's work the "first full-scale study in the English language of the rise of the machine in the modern world and one of the first scholarly studies in any language to emphasize the interplay of technology and the surrounding culture" ("The Myth of the Machine" 152). In contrast to materialist (Marxist) notions of culture, historical change according to Mumford was driven by alterations in human values, of which changes in technics were not the basis, but an outcome or expression (Williams, "Lewis Mumford as a Historian of Technology in *Technics and Civilization*" 63, 65).

After the disasters of the Second World War, which included the personal loss of his son and the increasing fear of nuclear obliteration, Mumford later returned to the topic of technics and culture on a grand scale. In a two-volume series entitled *The Myth of the Machine* he advanced a critique of the "megamachine," an abstract mechanized system of power, and sought to "question both the assumptions and the predictions upon which [the] commitment to the [. . .] forms of technical and scientific progress, treated as ends in themselves, have been based" (3).[50] These latter two books represent a continuation of Mumford's *telos* as an author-historian: to relate differently to accepted notions of a technological past. He did this by authoring histories that expanded the definition of technology beyond the mere development of tools to encompass the entire human endeavor of "mind-making, self-mastering, and self-designing" (9).[51] Mumford's works can thus be characterized as examples of a cultural approach to technology and a pivot away from a framework of necessity in history driven by technics. His approach was instead directed towards frameworks of at times personal, rational, and conscious, but also ephemeral and relatively arbitrary societal shaping of technology (Miller, "The Myth of the Machine" 162).

Susen's second key historical phase was concurrent with the rise of resistance through social movements starting around the mid-twentieth century. Susen underscores the role of political activity in changing perspectives of the intellectual tools used to create and transform common values. With regard to scholarly accounts of technology during this time, two books published in 1964, Leo Marx's *The Machine*

50 As Donald L.Miller summarizes, the myth of the machine is "the widely held view that technological progress and the expansion of power in its various forms—military, financial, and political—were the chief goals of the human endeavor" ("The Myth of the Machine" 152).

51 Susen bases his ideas that history is shaped by frameworks of contingency in critical thought on the assumption that historical developments are "lawless, unpredictable, nonlinear, directionless and particular" (138).

in the Garden, a classic text in American Studies,[52] and Marshall McLuhan's *Understanding Media*, a foundational text for Media Studies, provided very different "analyses of the societal effects of new technologies" (Benesch, "Does Technology Drive History?" 96).[53] One significant reason for this divergence is the fact that Marx's ideas for his book took form, as he describes in the afterword of the 2000 edition of *The Machine in the Garden*, in response to the events which occurred before, during, and in the immediate aftermath of World War II (367–372). After his studies at Harvard and time spent in the U.S. Navy, Marx worked "intermittently" on what had begun as a dissertation for over 15 years (369, 373). This protracted time and the fact that he continued to write on the topic of technology throughout his academic career, makes him a bridge figure between the first, second and third key historical phases and critical thought on technology with relation to the aforementioned cultural discourses. Of particular note for the current analysis is that Leo Marx's first book examines the subject of technology by looking not at historical, but at literary texts in the American tradition concerned with the "onset of industrialism in America" (367). Using various responses to sudden changes in nineteenth-century America life, Marx demonstrates how the intrusion of technology into the pastoral ideal is a reoccurring theme in the literary works he considered. His understanding of the relation between history and technology emphasizes the "reconceptualization and imaginative strategies" that helped Americans adapt new technologies into their "existing symbolic configurations" to integrate them into the "American experience" (Benesch 101). Analyzing the motif of what he calls the "machine in the garden" in literature, Marx spells out a more inclusive model of historical change that encompasses both technical and cultural aspects. That technology might drive history and historical change alone would be impossible for Leo Marx "without imaginative human interaction" (Benesch 101).

At the time of the book's publication in 1964, Marx supposed that this particular nineteenth-century imaginative interaction with technology was no longer valid "in the era of high technology," stating in his afterward to the 2000 edition that he had thought that "the pastoral response to industrial society almost cer-

52 Despite its status as a classic text in American Studies and obvious crossover in topic, *The Machine in the Garden* has not been included in the current book. While the machine as symbol is central to Leo Marx's analysis, his book is not a history of technology in a narrow sense. Marx has undoubtedly read Veblen, but only mentions him once in the book (307). What have been included in this work, due to their relevance to the current discussion, are Leo Marx's metahistorical observations on the subject of technology.

53 This and the following quotations from Klaus Benesch's article "Does Technology Drive History? McLuhan, Leo Marx und die Materialistische Medientheorie" are translations from the German by the author.

tainly had become anachronistic, and that it soon might be expected to lose its hold on the minds of disaffected Americans" (383–84). In fact, Marx admits that he had underestimated the power of the motif he analyzed to shape discourses around colonization by technology in the 1960s environmental and student movements.[54] His acknowledgment may only confirm the embattled notion advanced in his book that myths and symbols play a causal role in shaping material realities. But despite its marked tendency to universalize American culture, Marx's analysis does allow room for an interpretation of technology that is contingent on human and machine interaction. His book helps conceive of technology that is less determined, predicable, or linear, and more reflective of an interplay of material realities and historical discourses.

This trend toward a framework of contingency in the historiography of technology can also be observed in McLuhan's approach to causality in media environment. As will be demonstrated in the chapter dedicated to McLuhan, what may seem like a determinist framework is a misreading of McLuhan's understanding of formal causality. Recognized along with his senior colleague Harold Innis as the two "anchoring" members of the "Toronto School,"[55] McLuhan and Innis emerged as scholars of economy and technology from the edge of empire by reckoning with the colonial past (British) and colonial present (American).[56] This cultural proximity to power can and should be considered both a stimulus and a context for both authors' historical critiques.[57] In a personal letter to the President of the University of Toronto, McLuhan later acknowledged this line of thinking: "Canada as anti-environment to the U.S.A. is able to perceive many of the ground rules and operational effects of the

54 Two examples given by Leo Marx are the 1962 publishing of Rachel Carson's *Silent Spring*, which framed DDT and other biocides as the machine that intruded on harmonious life in the heart of America, and the Berkley student rebellion in 1964 that viewed the military industrial complex and the ongoing war in Vietnam as the machine that needed to be stopped through acts of civil disobedience (*Machine in the Garden* 380–84).

55 As Menahem Blondheim and Rita Watson write, the "Toronto School" is purely a retrospective and "invisible theoretical concept," which attempts to group a number of scholars under the umbrella of communications research (21–22). Additionally, they note that there is neither a consensus on the members of the network nor a unified theoretical stance that can be identified (Innis and McLuhan also had diverging views). For the purposes of this book, the fact that the school is mainly identified by its two core scholars and that it is geographically located at the edge of empires make it a representative choice for theories of technology in history (21–22).

56 In contrast to Americans Thorstein Veblen, Stuart Chase, Lewis Mumford, and Leo Marx, Innis and McLuhan were Canadian, which gave them a uniquely close, but outside perspective of U.S. culture.

57 Political Scientist Roland Deibert considered Canada's geographical positioning next to the U.S. as "*the* backdrop for all of Harold Innis' scholarship" (32).

American environment that are quite imperceptible to the U.S.A" (*Letters of Marshall McLuhan* 319).

Innis studied Canadian economic history. He had read Veblen and even compiled a bibliography of Veblen's works. He was particularly impressed by Veblen's ability to "outline the economics of dynamic change and to work out a theory . . . of dynamics" (Innis, "A Bibliography of Thorstein Veblen" 12). Innis's later work focused on the effects of communication in empire, and it is important to note that McLuhan's critique of media was modelled on Innis's economic analysis of staples.

McLuhan studied literature and popular culture, but his later research interests revolved around the influence of technology and communication in, not on, society. In his most famous book, *Understanding Media*, McLuhan associated technology with media and did not focus on the constituent aspects of technology, but instead on the physical effects that various media have on the human sensorium.[58] Written at a time when electronic media were in their infancy and TV was new in its dominant role as a medium of communication, McLuhan was mainly concerned with the "psychic and social adjustments" to the new electric environment that constitute the generally invisible "side-effects" of technology (McLuhan and Gordon 564).

In contrast to the other scholars, McLuhan's analysis has its origin in the individual and their ability to deal with altered ratios among the senses (McLuhan and Gordon xix). Yet, beyond this microscope of contingent effects in individuals, McLuhan seeks to broaden the scope of his theory of history to the macro, needing what Thomas Misa has deemed a "powerful all alluring abstraction" to serve as an agent of social change (320). Describing how mankind arrived at this point in history, McLuhan lays out four media cultures (oral/ear, literacy, print, electric) that follow from the dominant media of the time. As such, his seemingly deterministic history of technology, or rather historical narrative to understand media, has been described by Benesch as a "succession of linear progressive, interrelated media-technical revolutions" (97).

Despite this characterization of his overarching historical narrative, the current book argues that a framework of contingency concerning technology can be more differentially attributed to McLuhan. Neither Innis's early staples theory nor McLuhan's later notions about media attribute necessary causality to things (staples, modes of media), but rather attempt to show how these things contribute to form a constellation or an environment. In McLuhan's work, technological environments are not monocausal, but extremely dynamic and ever changing. They are not inevi-

58 A chapter-by-chapter analysis of different technologies makes up the second part of *Understanding Media* and includes both conventional media such as the press, radio, the telephone and television, but also many other unconventional media like numbers, clothing, housing, games, language, roads and money.

table and can be understood and controlled. In this respect, McLuhan's similarity to Veblen is that they both considered the potential to create practices and use technologies to emancipate humans in an evolving industrial/technological society. Admittedly, the connection between McLuhan and Veblen almost certainly must run through Innis. But such a comparison allows for at least a loose association to be drawn between the two. More central to this book is placing McLuhan within a larger trend towards frameworks of contingency in twentieth-century historical writing. For these men, finding meaning in a rapidly changing world meant penetrating the spaces of the past and filling them in with meanings of technology. The next chapter looks at the state of the arts at the beginning of the twentieth century.

Chapter 2
A Bad State of the Arts
Historical Accounts of Technology in Thorstein Veblen's Wake

It is one of the commonplaces of the received economic theory that work is irksome.
Thorstein Veblen, "The Instinct of Workmanship and the Irksomeness of Labor"

The knowledge of ways and means is a communal product.
Thorstein Veblen, *On the Nature of Capital*

If modern economic theorists were taxonomists, Veblen attempted the study of the embryology, morphology, physiology, ecology, and aetiology of economics.
Harold Innis, "A Bibliography of Thorstein Veblen"

A Framework for Technological Discourses

Born on 30 July 1857 in Wisconsin to immigrant parents, Thorstein Veblen lived through a period of massive economic and social change in the United States. Academically, Veblen studied and reflected on these changes and eventually wrote a sweeping narrative of societal critique for his time. He first received his doctorate in philosophy from Yale University in 1884. He accepted his first formal academic position at Cornell University in 1891 and then moved to become a reader at the University of Chicago in 1892. His first and most famous book, *The Theory of the Leisure Class*, published in 1899, espoused the notion that at the expense of basic needs, various economics classes also pursue what might be construed as frivolous goods and services, mainly because, for a sense of self-worth, they attempt to imitate the habits of the class immediately above them. Stuart Chase summarizes Veblen's famous theory of "conspicuous consumption" in more detail as follows:

> [Veblen] maintains that while the top group – the nobility, the aristocracy, or, in democratic countries, the plutocracy – sets the standard of conspicuous consumption and waste, the classes immediately below imitate this extravagance to the limit of their purses, in a very human desire to demonstrate to the world that they are as good as the next man. Each economic class tends to imitate the luxuries of the class immediately above it (Chase, *The Tragedy of Waste* 88).

Due to the popularity of *The Theory of the Leisure Class*, which is now a classic text of economic theory and is still in print, and his subsequent writings, Veblen can be considered one of the first important American economists and social theorists. His oeuvre is diverse and complex, and his breadth of subject matter

https://doi.org/10.1515/9783111396477-003

makes a general analysis of his theory too wide-ranging and beyond the scope of this chapter. The present discussion is limited to a targeted analysis of Veblen's framing of discourses about the concept of technology and an attempt to trace his legacy through subsequent scholars. These two points of interest were primarily stimulated by questions posed in two chapters (eight and nine) of Eric Schatzberg's monograph on the history of technology as a concept.

Building off the aforementioned scholarship, the following three sections each extricate a feature of Veblen's understanding of technology that emerge from his writings. The first feature is a critical account of the role of moneyed interests in the development of industry. The second is Veblen's adherence to what Schatzberg calls a "populist concept of technology."[59] And the third feature is Veblen's understanding of technology as a subordinate, but novel factor within a broader framework of human culture. This last point is significant in that it allowed future writers to elevate the concept of technology, which Veblen translated into English, to the status of a historiographical subject that served subsequent cultural-historical inquiries into effects in society.

In the section entitled "The Uses and Abuses of Veblen's Cultural Template for Discourses on Technology," the discussion turns to align the authors investigated in the subsequent chapters of this book with Veblen and the mold in which he critiqued understandings of technology. The remainder of the current chapter considers a few points of personal reference between Veblen and Chase, Veblen and Mumford, and then ends by reviewing an evaluation of Veblen's career and works by Harold Innis, who closely examined his writings. This last section underscores Veblen's contribution to what might be considered the dynamic study of the state of the industrial arts and its societal effects that others could emulate.

Savage Instincts – Domination of Technology by Moneyed Interests

Returning to a theme mentioned in the previous chapter, part of Veblen's thinking about the current state of the industrial arts at the turn of the twentieth century involved appropriating the discourses around the German word *Technik* into

59 Populist in the sense that technology should be non-elitist and not determined by those who currently owned the means of production or controlled the means of finance (Schatzberg, *Technology* 135). In other words, technology was intended for and should be experienced as a common and universal right by ordinary people.

English-speaking scholarship.[60] In addition to the changes to industry he observed in his studies of economics, reading German economic historians from Karl Marx to Gustav Schmoller to Werner Sombart must have convinced him "to incorporate the concept of technology into his theory of history" (Schatzberg, *Technology 125*). As Schatzberg explains, to serve as a key concept in his theorizing about economic and societal change, the Norwegian-American polyglot "introduced a new meaning of technology into academic English," that of "industrial Technik," which combined the "industrial arts in general" with the "material means of production" (118, 136).

It would have been easy to overreach and employ such a general concept to help explain historical events and other cultural or social changes. Indeed, the concept of technology was a significant factor in Veblen's Darwinian-inspired evolutionary-social theory, but following Sombart's example, he did not view technology as a driving force in history.[61] Although Veblen regarded the industrial system to be a material framework for society, he considered modern business enterprise – modern capitalism – to play a much more dominant role in shaping how technology, and consequently society, was transformed. In Veblen's assessment, technology could be culturally and collectively beneficial or used in a parasitic manner. He differentiated primarily between productive evolutionary impulses for society, like "the instinct of workmanship," and unproductive ones within a community, like "predation." Derived from this universal dichotomy of human behaviors,[62] he believed that technology could not reach its full cultural potential due to the manner in which business enterprise could make decisions not to fully support functionality and innovation, but to seek "pecuniary gain" and the "accumulation of wealth" (*The Theory of Business Enterprise* 20).

In *The Theory of Business Enterprise*, one of Veblen's early works, a link between technology and history is so strong that it borders on a conflation of the two. In this sense, technology could be assumed to have an epochal and progressive meaning. The word referred to a new situation, a "different state of the industrial arts" characterized by science in the service of technology, at this point understood as the study of techniques (302).[63] New capabilities set free by increased power,

60 Clear historical evidence of a direct link between the German discourses of *Technik* and a new American discourse of technology can be found in Veblen's 1903 review of Werner Sombart's *Der Moderne Kapitalismus* (Schatzberg, *Technology* 129).

61 The first chapter of the second volume of Werner Sombart's *Der Moderne Kapitalismus* explicitly rejects *Technik* as a driving force in history (1–24).

62 See Veblen's *Instinct of Workmanship* (27–28).

63 As in the phrase "technology of the production process." On the same subject, see also Werner Sombart's "Die Neue Technik" (*Der Moderne Kapitalismus* 67).

standardization, and efficiency could emancipate humans to pursue less work, and ultimately to improve their conditions of life. This might hold true, according to Veblen's thought, if the capitalist system, along with its institutions, habits of mind, and wasteful consumption did not often contradict developments in industry. In his interpretation, the economic system had become a means for profit-seeking financial institutions to reward "absentee ownership" (Veblen's term to indicate disinterested owners separated from day-to-day operations). Control of technical knowledge and production in industry needed to be in the hands of those with deep technical knowledge. "In every-day phrase," as Veblen stated, "under the rule of the current technology and business principles, industry is managed by businessmen for business ends, not by technological experts or for the material advantage of the community" (*The Instinct of Workmanship and the State of the Industrial Arts* 351). He characterized this dispossession of technical knowledge from the ideal of collective ownership to the reality of private profit as the "savage state of the industrial arts" (103–37). It is from this nonconformist characterization of technology's status that others who read and admired Veblen's analysis would ultimately begin their own historical inquiries.

A State-of-the-Art for the Wealth and Well-Being of the Many, Not for the "Timeless Gain" of the Few

Veblen historicized the concept of technology with a critical eye toward its relation to business enterprise (here meaning American capitalism).[64] He adopted a fundamental theoretical separation between industry and business, which allowed him to critically analyze the economic system and culture in which American technology emerged. Because Veblen thought that industry should be directed toward the wealth and well-being of the community, he recognized that under ideal conditions, technology would be considered as part of a common stock of knowledge and used in relation to a flourishing communal culture. It follows that he supported the productive and collective aspects of technology, especially engineers, whom he considered inclined to create and build a better state of the arts.

To Veblen's discontent, however, the reality he observed having lived through the Gilded Age, the time he was reacting to in his earlier writings, was that this

64 Veblen criticized a behavior this timeless gain was thought to encourage, namely demonstrating one's status through conspicuous consumption. Veblen's "*Angriff auf die Kultur*," or attack on culture, to quote Adorno, was an attempt to discredit this practice as barbaric (Adorno 82–111). Veblen's analysis of technology focused more on the instinct of workmanship, the industrial arts, the material means of production and their relation to business enterprise.

period remained characterized by extreme exploitation on one hand and gluttonous behavior on the other. Particularly in the United States, industry rarely belonged to the community and was frequently in possession of private owners and financial institutions or other businessmen. As a consequence, ownership of the means of production allowed the so-called captains of industry, or as Veblen called them "pecuniary magnates," to accrue and hoard mass sums of capital. Any further financial gains from this capital Veblen described as being "timeless," meaning that the gains stood in no relation to time and were disassociated from "technological circumstances and processes of industry" ("On the Nature of Capital" 125–29). In short, the perpetuation of business practices divorced from the material means of production and the knowledge of industrial processes stood to financially benefit the few, not the many.

For his part in advancing a dissenting view of technology's status (the state of the industrial arts), Veblen was and still is largely marginalized within academia. Yet by adhering to a more universal and populist concept of technology (the industrial arts in general, including the means of production), which he recognized as being "rooted in human agency, with skilled workers and production engineers at its vanguard," Veblen created a mold in which subsequent North American writers could map out relations to a technological past (Schatzberg, *Technology* 135). Although Eric Schatzberg is pessimistic that Veblen's populist concept of technology survived World War I, the current book finds value in the historical and cultural framework that Veblen provided his readers for subsequent accounts of technology.

A Novel Factor in the History of Human Culture

Part of the theoretical groundwork laid by Veblen was being able to distinguish the concept of technology from the culture and logic that animated and structured its industrial application at a given time. He demonstrated this by tracing the origins of a loss of habituation in a specific cultural practice that largely took root in England, and not until later in Germany and continental Europe. Before World War One broke out, Veblen had started working on an essay later published in book form as *Imperial Germany and the Industrial Revolution*, about the historical circumstances that had shaped technological development in Germany and in English-speaking countries. Although he considered the German and English roots of industrial development as "two distinct and somewhat divergent lines of the cultural development in modern times," Veblen stuck to his more universal notion of technology (Preface). Part of this notion was not assigning causality for Germany's industrial development to the "logic of manifest destiny, providential nepotism, national genius, and the like," but rather to the contingency of adopting a practice that had developed over generations which formed the basis of the "machine technology" (182–88). In a passage worth quoting at length, Veblen described the origins of the ultimate and fundamental shift in the ways of thinking in pre-industrial England that underpinned this concept of technology:

> The epoch-making intellectual achievement of the English technologists and their like, who prepared the Industrial Revolution and have afterward worked out its consequences in technology and the material sciences, is not so much that they gained a new manner of insight into the nature and working of material things, as that they were, by force of circumstances, enabled to forget much of what was known before their time; by atrophy of the habitual bent for imputing anthropomorphic qualities and characters to the things they saw, they were enabled to interpret these things in terms of matter-of-fact (183).

Veblen illustrated that the novelty of the shift could be observed not in innovation of methods, but rather in moving away from unscientific ways of thinking. He continued, citing the general historical circumstances that resulted in the rise of universal habits of thought, first in England, and then their rapid transfer and subsequent adoption on the European continent:

> Circumstances, the chief of which appears to have been the decay of personal government consequent on the decay of dynastic enterprise, weakened the inveterate habit received from the past of construing material phenomena in occult, magical, quasi-personal, spiritual terms. This habitual imputation of spiritual potencies and correlations among material facts had fallen into abeyance in the Island [England] sooner and farther than on the Continent, and the curious ones among its inhabitants therefore fell more readily than their Continental neighbors into the prosaic habit of taking these external things at their face value, as opaque matter of fact, and so construing them and their movements in terms of (relatively) unsophisticated sense perception, without metaphysical afterthought, or with a negligible residue of such afterthought (183–184).

Veblen pointed out that the rise of merchant classes destabilizes the monopoly of monarchs and religious leaders to influence popular imagination. These are the conditions that allow for the emergence of technological thought, which can be broadly understood as a scientific way of thinking. Veblen then explicitly stated his universal and populist concept of technology: "In their elements, therefore, the premises and logic of the machine technology are in every man's mind, although they may often be overlaid with a practically impermeable crust of habits of thought of a different and alien sort" (*Imperial Germany and the Industrial Revolution* 183–84).

Given Veblen's interpretation of technology that privileged an ordinary person's ability to possess techno-logic, in a book written during World War I, *An Inquiry into the Nature of Peace and the Terms of its Perpetuation* (1917), he later questioned the mechanistic application of the dominant industrial logic and the economic order upon which nations depended. Questions of war and peace would be decided by how technology was applied and, with regard to Veblen's concept of technology, how technological, as opposed to patriotic, peoples and their leaders would become: "This modern technology, with its underlying material sciences, is a novel factor in the history of human culture, in that addiction to its use conduces to the decay of militant patriotism, at the same time that its employment so greatly enhances the warlike efficiency of even a pacific people . . . " (314). The appearance of the words use and employment in this passage – in the same sentence – is rather telling. These two nouns both refer to actions that can be put into effect through technology. Their meanings express a subtle difference and one that is worth rereading: technology's use (use of the techno-logic as a form of knowledge) theoretically has the effect of reducing nationalism, while technology's employment (instrumental application thereof) for purposes of efficiency in war is problematic.

The above analysis demonstrates the particular manner in which Veblen framed discourses around technology for an English-speaking audience. He separated business from industry. This dichotomy allows him to conceive of technology, not as a private or exclusive entity, but rather as a way of thinking scientifically that can be easily taught and learned by different societies. Understood to involve a common stock of knowledge for the benefit of society, Veblen's popular and culturally embedded concept stood in contrast to the employment, on a national and political level, of both mechanized weaponry for state warfare and nationalistic propaganda for mass mobilization.[65] By conceiving of technology as a "creative expression of human culture," Veblen created a template by which others could study technological effects in the modern industrial era (Schatzberg, *Technology* 3).

---

65 The destruction of industrial warfare was enhanced by mass-scale production of weapons and improved communication and transport technologies during World War I. The Committee on Public Information, or Creel Committee, was established under the Wilson administration to whip up public opinion in the United States to join the war effort. It is worth questioning the common interest in producing mass weaponry that kill and systems of propaganda that deceive. As John P. Diggins writes, "George Creel tried to exploit the book's anti-German arguments; but the Post Office, perhaps uneasy about Veblen's harsh treatment of the British and his depiction of the Prussian business imagination as superior to the American, barred the book from the mails as pro-German" (189).

The Uses and Abuses of Veblen's Cultural Template for Discourses on Technology

Connecting to a discussion advanced in chapter nine of his monograph *Technology – Critical History of a Concept*, Schatzberg frames Veblen's legacy as a dichotomy between cultural and determinist understandings of technology. The cultural view of technology is considered to be "imbued with human values and strivings in all their contradictory complexity" and "express the spirit of an age," while the determinist understanding employs an instrumentalist view in which technology is a "mere instrument that serves ends defined by others" and is an "autonomist, largely beneficial agent of social change" that is associated with the idea of progress (Schatzberg, *Technology* 3, 136). In Schatzberg's book, a determinist group of scholars is represented through examples like economist Herbert Davenport and historian Charles Beard, and by the so-called Technocracy Movement of the late 1920s and early 1930s. To contrast this determinist camp, Schatzberg claims a cultural view is best epitomized by Lewis Mumford.

Taken together, these two groups represent the uses and abuses of Veblen's cultural template for discourses on technology. Abuse of the template has led to narratives of necessity and technological determinism, while uses, in particular by Stuart Chase and Lewis Mumford, have helped expose the underlying incompatibilities between systems of technology and the well-being of a plurality of people in society. The next few chapters of the current book represent an attempt to expand on the latter line of analysis by examining scholars, two of which who were directly influenced by Veblen,[66] who wrote historical accounts of technology with similar conceptual concerns, such as the direction of American civilization or the well-being of all classes in U.S.-American society. Three of these scholars, Stuart Chase, Lewis Mumford, and Marshall McLuhan, are identified as having explored cultural understandings of technology after Veblen. While Chase and Mumford are analyzed in the context of their responses to the mechanization and industrialization of U.S.-American society and culture, the interpretation of McLuhan offered in the current study is an attempt to show the continuation of a trend toward contingency in historiography further removed from Veblen and into the second key historical phase, which anticipates the emergence of postmodern thought. Although Veblen does not figure prominently in McLuhan's writings, as will be shown later in this chapter, Veblen did directly influence his colleague in Toronto, Harold Innis. This is significant because after Innis shifted research areas from economics to communications, as will be shown later, it was McLuhan

66 Influence deemed to be substantial mainly through citation of Veblen's written works.

that adopted Innis's historical framework and methodology and who attempted to expand on his line of research. This was necessary, as McLuhan observed, because he thought that Innis "lacked training in the arts" (Gordon, "McLuhan's Compass" x). But before more in-depth analysis of the authors is presented, the rest of the current chapter will be used to briefly reveal the connections that link Chase, Mumford, and McLuhan to Veblen.

Stuart Chase

Thorstein Veblen had a profound effect on Stuart Chase. As someone interested in the economic system in the United States, Chase greatly admired Veblen and modelled his writings on the cultural template that Veblen had provided. Several points of evidence support these claims. First, Veblen is by far the most referenced economist in Chase's writings. Among the myriad of references and sources in his books and pamphlets, quotes are taken from a number of Veblen's books which include but are not limited to *The Theory of the Leisure Class*, *The Theory of Business Enterprise*, *Absentee Ownership*, and *The Engineers and the Price System*. Second, in extolling praise on Veblen, Chase used superlatives. He went so far as to venerate Veblen as "the greatest economist whom America has produced" (*Technocracy* 5). Other laudatory remarks can be gleaned from the foreword to the 1934 edition of *The Theory of the Leisure Class*, in which Chase refers to Veblen as one of his idols when he was a young man. Although he would eventually attend some of the same conferences that Veblen did, Chase had never spoken to the elder professor in any depth. This situation changed after Chase sent Veblen a copy of his first book, *The Tragedy of Waste* (1925), to which Veblen responded by personally inviting Chase to a dinner. After their meeting, Chase recounts from memory what one of the "wisest Americans of his generation" had to say about his first book:

> Your book is too simple, too logical. Man is not a logical animal, particularly in his economic activity. But you have gathered some interesting facts. You have documented the conception of economic waste more thoroughly than it has been done before. But, do not expect any attention to be paid to it. Work of this nature, excellent as it may be, does not command attention, at least not yet. Some day, perhaps, I wanted to tell you that I appreciate your references to me, and that I think your work, while oversimplified is a contribution to that kind of economic thought in which we both are interested. You have a feeling for facts. Hold to them . . . (*xi–xii*).

Later in the foreword, Chase affirms Veblen's success in gathering facts and synthesizing them into "one of the boldest interpretations in the history of economics" (xiii). Such lofty words represent high praise for Veblen's book, but it is evident from the quote above that Veblen also recognized value in Chase's work. Indeed, Veblen seems to confirm common ground and interest both in the type of thought they both pursued and the type of facts that Chase presented in his first book. Could this sentiment also serve as approval of his framework to study economics? It is hard to determine based on this quote alone, but it is certainly suggestive. Finally, a piece of direct evidence that Chase modelled his writings on the cultural template that Veblen had provided was a closing comment in the forward to the 1934 edition. In Chase's opinion, the use of Veblen's template for discourses on technology took precedence over future forecasting or economic outlook. Even more important than the conjectures that he predicted or misjudged, Chase lauds Veblen for foretelling the "mold into which that history must fall for decades to come" (*xiv*). This statement seems to hint at the reluctance to place historical outcomes into a framework of necessity. Stuart Chase's use of Veblen's critical mold for discourses about technology will be explored more closely in the next chapter, but it seemingly culminated with a rarely discussed book that Chase had published in 1929. Encouraged by Veblen's chastising but instructive remarks, Chase constructed an entire monographic study around the supposed social effects of technology entitled *Men and Machines*.

Lewis Mumford

To what degree did Lewis Mumford profit off of Veblen's groundwork with regard to discourses about technology? Mumford was critical of Veblen's ultimate response to the state of the industrial arts, what he called a "practicable soviet of technicians" in *The Engineers and the Price System* (Veblen, 138). Yet Mumford can be shown to draw on the Veblen template to launch his own critique of technology and to detail the changes he observed in society over a long period of development since the Middle Ages. Evidence for this claim can be taken from Mumford's writings in the form of personal anecdotes and direct acknowledgements that will be elaborated in more depth in a later chapter.

"Before meeting Veblen," Mumford wrote in an autobiography of his early years, "I had read all of his books and had been stirred by them" for their originality and "cultural range" (*Sketches from Life* 220). Mumford would eventually study and work with Veblen. He first encountered Veblen in person at the *New School for Social Research* in New York. The young New Yorker recounted his excitement at the prospect of learning from the established Midwestern professor: ". . . another private dream of mine came true; for that spring [1919] I took a course in modern economic development under the scholar I had already admired from a distance . . ." (*Sketches from Life* 212).

After a short stint working together at *The Dial*, a New York magazine that published political and literary criticism, Mumford would begin his own career as a writer. In his first monograph dealing explicitly with what is now called technology, *Technics and Civilization* (1934), Mumford gave credit in the works cited section to Veblen's distinctness as one of the few Americans to write authoritatively on what he at the time called technics: "Thorstein Veblen's works, taken as a whole, including his less-appreciated books like *Imperial Germany* and *The Nature of Peace*, form a unique contribution to the subject" (448). Later in the bibliography of *Technics and Civilization*, Mumford reiterated the previous remark and followed it with a comment on the cultural adaption needed to adjust to a dynamic technological environment:

> After Marx, Veblen shares with Sombart the distinction of being perhaps the foremost sociological economist. His various works, taken together, form a unique contribution to the theory of modern technics. Perhaps the most important from the standpoint of technics are *The Theory of Business Enterprise* and *Imperial Germany and the Industrial Revolution*: but there are valuable sections in *The Theory of Leisure Class* and in *The Instinct of Workmanship*. While a believer in rationalized industry, Veblen did not regard adaption as the passive adjustment of an organism to an inflexible physical and mechanical environment (472).

When Mumford had to teach a course at Columbia University from 1932–1935, he realized that other than Veblen, there was not much American literature on technology.[67] While he was able to engage in his own research agenda by directly visiting the technical museums of Europe and reading European sources on the subject, Veblen had provided Mumford a framework to historicize the contingencies of technology in an American setting. As an "economist well-grounded in technics and anthropology," Mumford appreciated Veblen's "unconventional approach, his insidious criticisms of the business establishment" and his "readiness to speculate on matters where more positive evidence was . . . just – with Veblen's help! – being discovered!" Despite Veblen's extensive and innovative "historic scholarship," Mumford does criticize him for "pinning his faith on a chimerical 'soviet of engineers' as holding the key to both economic productivity and social justice" (*Sketches from Life* 220–221). But as noted above, this criticism was about Veblen's answer to the state of the arts, not his historical framing of technology.

67 "Except for the work of Thorstein Veblen, the American literature of technics was singularly sparse" (Mumford, Lewis, *Sketches from Life* 449).

Harold Innis

As a Canadian economist that studied in the United States, Harold Innis represents a bridge figure who transferred knowledge between discourses on technology from Veblen to Canada. He did this by summarizing and commenting on a range of Veblen's works in the form of a general and overarching analysis of Veblen's thought. Innis's authoritative "A Bibliography of Thorstein Veblen" (1929) strongly suggests that he thoroughly read and closely studied Veblen's books and articles. Innis's thoughts and observations on Veblen's oeuvre also provide insight into the features of Veblen's analysis that Innis deemed valuable for his own research.

To explain Veblen's work, Innis began his summary by considering how Veblen had been influenced by his upbringing and educational environment.[68] Innis recognized that Veblen had been affected by the legacy of post-Civil War America and the rise of American industrialism. He also notes Veblen's connection to the German economic historians and his approach to the subject of economics – a philosophical background that involved inductive reasoning and an interest in dynamics (3–4).

Having provided this background to Veblen's early influences, Innis observed that the content, but also the form of Veblen's economic analyses should be emulated, with one caveat. Initially, Innis commented that Veblen's early articles were "models of analysis of economic facts" and that they could be used as "models for the work of later economists in analyzing the relationship of complex factors" (3–4). But when important contributions were published at the turn of the century, Innis saw the style in which Veblen had written his reactions to the Gilded Age as "unfortunate." This was because his use of phrases like "conspicuous consumption" and "pecuniary emulation" had stuck. For this, Innis claimed that Veblen's work never recovered because he was characterized as a satirist (7).

With regard to interpretations of Veblen's later work, Innis referred to Veblen's most important contribution to the philosophical basis of economic theory was attempting to create an alternative to motivation based on Jeremy Bentham's notion of utilitarianism by attacking economics on both the production and consumption side (7).[69] In other words, Veblen had argued for economics not based in hedonism of the individual, but rather based on functionality in production

68 The bibliography was "intended to stress the importance of the environmental factor" in Veblen's work. Innis wrote that, "having actually lived through one of the economic storms of new countries, he [Veblen] attempted to work out some of their important characteristics. His contribution to economics is directly in relation to this background" ("A Bibliography of Thorstein Veblen" 13).

69 Jeremy Bentham (1748–1832), considered the father of modern utilitarianism.

and use for the community. This theme, a concern for functionality in production as opposed to waste, was adopted by Stuart Chase in his book *The Tragedy of Waste*.

In one sentence, Innis characterized Veblen's main argument in books published after 1914 as confronting the growing gap between increased productivity in industry and the purposeful use of technologies, "machine industry was overwhelmingly and increasingly productive, and that the problems of the machine industry were incidental to the disposal of the product" (9). This analysis was followed by the Canadian scholar's effort to divide Veblen's work into constructive and destructive elements. Innis wrote that the constructive part of Veblen's work was his analysis of the effects of machine industry and the industrial revolution, and a critical assessment of the state of the industrial arts, which he thought needed reorienting. Conversely, Innis described the destructive part of his work as both modest and mostly limited to articles and reviews that Veblen authored toward the end of his career. These articles' main concern was with economic theory of the time in opposition to classical and neoclassical economists and support for the German Economic Historians (9). Innis thought that much of Veblen's constructive analysis of the effects of industrial society had been somewhat lost amongst discussions of theoretical criticism.

Despite some negative reviews in the wake of Veblen's scholarship, Innis remained intrigued by the American professor's methodology. Innis lauded Veblen's approach by observing that in contrast to modern economic theorists that approached economic study through classification like a taxonomist, Veblen attempted to study the entire life cycle of economic institutions or the "embryology, morphology, physiology, ecology, and aetiology of economics" (10). Innis was so impressed with Veblen's approach to the study of economics he even dismissed the importance of Vebeln's research results, writing ". . . it is the method of approach which must be stressed, and not the final conclusions" (10). Primarily cautious, analytical, and objective throughout the bibliographical review, Innis made a significant comparison toward the end of the piece. Innis drew a parallel between Veblen's position at the end of the nineteenth century and Adam Smith's two-hundred years previously. Both scholars made unique and timely contributions to economic history: Adam Smith's stock-taking at the end of mercantilism and the beginning of the industrial revolution and Veblen being among the first to analyze tendencies in a society with machine industry.

To conclude his analysis, Innis emphasized both the ever-changing subject of Veblen's work, studying processes of growth and decline of institutions and associations. This approach helped Veblen discover economic trends in order to escape or avoid their effects (Innis, Harold 10).[70] What emerges from Innis's assessment is the importance of the approach, the template that Veblen had created for what Innis referred to as a "constructive warfare of emancipation" from dominant frameworks to understand technology that are unable or unwilling to survey the cyclone of societal effects that have resulted from dynamics in the economy since the Industrial Revolution (11).

Veblen died in August of 1929. The following months witnessed the "great" Wall Street stock market crash of 1929. Veblen was proven right to question the business enterprise and expose those responsible for the savage state of the industrial arts. The task to frame technology within a common cultural history, or as Innis formulated it, to liberate scholarship from the struggle between "Frankenstein's monster and Frankenstein" was left to be taken up by other authors (11).

---

70 The importance of understanding effects, as Veblen pointed out, could be observed in that "England, although the first to feel the effects of the industrial revolution, has never been conquered by it. America and the new continental countries have been less fortunate" ("A Bibliography of Thorstein Veblen" 11).

Chapter 3
Disequilibrium, Unwilling Robots, and Decreation

Stuart Chase's Economic – Functional – Political Relations to Technology

The dilemma in its simplest terms is that the credit system has not kept step with the technical arts.
Stuart Chase, "The Enemy of Prosperity"

The two great inventions for saving human labor are money and the machine. Both have got out of hand, and are prancing and trampling around through the social order as though they had equestrian rights from God.
Stuart Chase, *A New Deal*

The price of this bright Utopia [a power system based on electricity] is an economic system which is based on use and function, not on debt and profit.
Stuart Chase, *The Promise of Power*

A Man Between Theory and Practice

Stuart Chase is an interesting character to study for the relationship between technology and society in the early twentieth century because he occupied a unique space between theory and practice. As will be shown in the following two chapters, Lewis Mumford and Marshall McLuhan represent more theoretical attempts to negotiate a relation to the technological past in order to orient readers in the present. In contrast, Chase wrote fast and for a broad audience. He incorporated just enough history and theory into his writings to be taken seriously. An anecdotal, easy-to-read style combined with statistical evidence won him the ability to influence economic and industrial policy and policy makers at the highest levels of government.[71]

Typically, Chase has been characterized as a popularizer of Veblen's ideas about technology and society. As historian Robert Westbrook criticized, Chase is

71 Chase corresponded with and advised many office holders that include congressmen, senators, and even U.S. president Franklin Delano Roosevelt.

Note on chapter title: Chase sought to evoke the past for current use or function to benefit society, as in his employment of R.H. Tawney's idea of a functional society.

https://doi.org/10.1515/9783111396477-004

less historically significant for the "profundity of his ideas" or for his originality, than for the place he occupied in the popular imagination of progressive critiques of American capitalism in the late 1920s and early 1930s (389). According to Westbrook, Chase's publications functioned as a "collection and distribution point for theory and research" for what has been identified as group of technocratic progressives.[72] Broadly defined, classification of Chase as a technocratic progressive means that one of his goals was to help make "technological civilization function more efficiently on its own terms . . . by a transfer of scientific and technical rationality to economics and politics" (Jamison 86). After a stint in government, he attempted to achieve this goal through his writings. Chase's main interest was not money. His personal correspondence reveals that he turned down a number of business opportunities to stick to writing. This leads one to believe that he must have strongly believed in utilizing the great potential he envisioned for technology to be used in the public interest (Westbrook 388). Possibly because of this utopian ideal, his books enjoyed wide readership, which allowed Chase to become a kind of mediator and later advisor for rational thinking about technology and the economy. The term rational, as ominous as it may appear to contemporary readers, refers to Chase's pre-1930s, largely socialist-inspired, non-fascist philosophy to apply matter of fact, engineer-like thinking to serve humans' needs and address human problems. In Chase's view, capitalism in its then current form had underserved society. He was primarily concerned with the purported effects of technology on the quality of life in the United States. His wide-ranging readings of diverse literary and technical literature and subsequent early books and pamphlets reflect this dominant theme.

Consideration of Chase's numerous handwritten notecards at the Library of Congress reveals that he was a sophisticated reader who questioned more than the technical rationality of capitalism. According to his notes, Chase was well aware that capitalism "welcomed technology and exploited it in detail," but blocked its full realization or even harmed consumers in pursuit of profit. His writings on technology, seldom referenced, offered his readers and sympathizers more than the choice of an "exchange of masters," or a switch in control of economic planning from capitalists to engineer-experts (Westbrook 408). Because Chase's early writings are not well-studied, this chapter presents Chase's analysis mainly through a review and analysis of two of his books on technology – *The Tragedy of Waste* (1925) and *Men and Machines* (1929). In these publications,

72 According to Westbrook, this group includes, but is not limited to: Thurman Arnold, Charles Beard, Adolph Berle, Bruce Bliven, Stuart Chase, Morris L. Cooke, Paul Douglas, Freda Kirchway, Max Lerner, Harold Loeb, Lewis Lorwin, Robert M. Lovett, Robert Lynd, Gardiner Means, George Soule, and Rexford Tugwell (Westbrook 388).

Chase attempts to question notions of industry in the United States through a study of industrial waste and a survey of the supposed effects of technology in society. In addition to his criticisms of production in industry, Chase was also one of the first to advocate for an early form of consumer protection. His 1927 book *Your Money's Worth*, co-authored with Frederick Schlink, helped start the consumer movement. Ultimately, in opposition to Westbrook's characterization of Chase's intentions, this chapter provides evidence that Chase actually distances himself from the "commercial and material formula of wealth and happiness" ("Declaration of Independence" 30) supported by the business community by declaring his independence from what Westbrook termed the "capitalist's identification of the good with the goods" (Westbrook 408).

Pinning his approach to technology to a "rationalistic view of life," which first pursues basic needs,[73] what emerges not only from Chase's early writings, but also from his work in government and subsequent dismissal, and from his choice to pursue a career as an author of popular books is the narrative of a man that clearly understood the complex effects of the technological system – including the negative ones – and consequently promoted a reconfiguration of technology's role in modern industrial society. Following Veblen, Chase thought of technology as subservient to human desires and in service of human needs. For this and other reasons detailed below, Chase's early accounts of technology can be understood to be in line with the other authors analyzed in this book who argued against an autonomous and determinist understanding of technological change. According to the arguments presented in this chapter, Chase hoped to convince his readers to "face the full implications" of technology to gain popular, more expert control of a technological and industrial order that had "run amuck" (*Men and Machines* 347–348).

Engineering and Accounting Background

Chase was born in 1888 in Connecticut to a family that could claim ten generations in New England.[74] His father, Harvey Stuart Chase, had a famous accounting firm

73 In his journal pages dated 1913, Chase supported a "rationalistic view of life." According to his notes, the aim of life is human service, which is made up of "practical factors" that include fundamental necessities like 1) food 2) shelter 3) clothing 4) education 5) health 6) legitimate recreation 7) government and law 8) finance 9) scientific exploration and 10) art ("The Rationalistic View of Life").

74 According to an ancestral card in the Stuart Chase archive of the Library of Congress, Aquila Chase, born in 1618 in Chesham, England, was the first member of the family to immigrate to the United States. Stuart's father, Harvey Stuart Chase was written in pencil as the ninth generation

in Boston and, among various other official publications, in 1902 he wrote a public report for the mayor on the financial conditions of the city. The reputation he gained by creating systems of uniform reporting and accounting for public entities led to him being tapped by President Taft to be a member of the Commission on Economy and Efficiency in the early 1910s (Westbrook 389). The younger Chase, meanwhile, had first attended the Massachusetts Institute of Technology to pursue a technical education, but then transferred to and graduated from Harvard University in 1910 with a degree in economics and accounting. After becoming a certified public accountant and working for several years at his father's firm, he took a government job for the Federal Trade Commission in 1917, where he was tasked with controlling food prices as part of rationing policy during World War I. More specifically, under the Food Administration, he was in charge of regulating profits of the meat-packing industry. In 1920, after an investigation, Chase authored a report entitled "Part V – Profits of the Packers" that found accounting irregularities in the packing sector. Unfortunately for Chase, the vested interests in the industry and their allies in Congress made him an enemy. Every attempt was made to target and undermine Chase's authority as a person and to downplay the report's findings by accusing him of socialist sympathies. For example, he was smeared by Indiana Republican Senator James Watson, who framed Chase's political beliefs as corrupt and anti-business. Watson accused Chase of associating with other extreme Socialists, radicals and anarchists and called him a "well-known exponent of Socialistic doctrines" and the "president and organizer of the Fabian Club, a society founded for the express purpose of furthering the doctrines of Socialism."[75] Sidestepping any wrongdoing on the part of the industry, the Senator focused his ire on the messenger of potential crimes, Chase, contending that men who are working for the Federal Government should be "investigators, not persecutors." If being attacked in the press and labelled a red had not discouraged him enough, Chase was later relieved of his job because of "limited appropriations" at the end of 1920.[76]

---

stemming from Aquila. That makes Stuart chase the tenth-generation Chase in New England. The Papers of Stuart Chase. Manuscript Division. Box 13, Subject File Miscellany 1931–1932, 1941–1944, 1951–1978, n.d. Library of Congress. Washington D.C. 2 January 2020.

75 These quotes were taken from a newspaper article titled "Reds hold government office, Watson claims" archived in The Papers of Stuart Chase. Manuscript Division. Box 2, Subject File Chicago Meat Packers Federal Trade Commission "red charges" 1919. Library of Congress. Washington D.C. 2 January 2020.

76 Quote from the termination letter that Chase received from the Federal Trade Commission, archived in The Papers of Stuart Chase. Manuscript Division. Box 2, Subject File Chicago Meat Packers Federal Trade Commission "red charges" 1919. Library of Congress. Washington D.C. 2 January 2020.

Disaffected, but wiser as a result of his experience as a controller, Chase did not give up trying to find solutions to the domination of industry by self-serving business interests and their political henchmen. He had been reading Veblen, who at the time (1921) was living in New York and teaching at the New School for Social Research. Still in Washington, Chase was asked to join the Technical Alliance, centered in New York. According to Chase, this organization was formed "to give concrete expression to Veblen's challenge" of letting engineers, as opposed to business enterprise and vested interests, run the economic system (*Technocracy: An Interpretation* 7). Chase moved to New York and became a member but, as a formal group, the Technical Alliance did not last long. Most of the members went separate ways at the end of 1921. Soon after, in 1922, Chase was with the Labor Bureau, which did research, accounting, and other professional services for labor groups and cooperatives. During this time, Chase began to study and write about the problem of "waste" in American industry. Combining the practice of statistics and balance sheets from his accounting background with an engineering perspective acquired from reading technical literature, he penned his first book in popular economics, *The Tragedy of Waste*, which was published in 1925.

Improvement of the Human Condition as an Engineering Challenge

Chase's general concern in *The Tragedy of Waste* is to illustrate the waste and lack of efficiency in industry in 1920s America. Having witnessed a remarkable increase in "the technical arts of production" in the last thirty years, Chase notes that while the quantity of goods and services has certainly increased, the standard of living for the average family in the United States has not kept pace (12). This assessment leads him to ask what benefits in "livelihood, comfort, leisure, and beauty" normal families incur from an increase in machines and power (13). The discrepancy between technological advance and lack of quality in the improvement of peoples' lives Chase attributes to waste. For the purposes of his book, Chase uses an engineer's definition of waste, described as the "maximum use value of output at minimum real cost in energy and materials" (16). This definition also incorporates notions of efficiency by adopting the engineer's goal of accomplishing "useful function with a minimum expenditure of energy" (25). As a consequence of these characterizations, the tragedy of waste he is referring to is the inability to take full advantage of the potential the technical arts have to improve the quality of life, even in modest terms for basic needs.

The following analysis will demonstrate that the uniqueness of Chase's contribution to early discourses on technology was to map out and develop a theoretical discussion about the downsides of technological change and to suggest alternate ownership and control of its application toward broader human needs. In his argument, Chase follows the Veblenian distinction between business and industry, or between an economy based mainly on acquisition and controlled by investment bankers and one oriented more toward a functional role for the technical arts in society. Far from advocating for a narrow perspective on human needs with instrumental control of technology in the sole hands of technocratic experts, Chase wanted to fundamentally change the way in which one relates to technology in society. He does this, for example, by including the rights to education and health care and his critique of wastes in natural resources into his analysis. He also reminds his readers that, in a functional society, technical knowledge ought to be free. Under the current economic system, he noted that society only shared in scientific discoveries to a degree and only after a period of "maximum obstruction" (204). In other words, in a system geared toward profit, suppression of new inventions served the vested interests because it prevented competition.

Chase uses quantitative methods of accounting to underscore his argument against waste. These measures are not based on monetary values, because they can tend to fluctuate, but on the "physical counts" of man-power and raw materials (38). Using statistics compiled and supplied by the Labor Bureau where he also worked, Chase set out to provide the quantitative data to support one part of his argument, that something must be done to better manage industry on a national level. As an example of the possibilities of industrial control on a national scale he cites the government policies implemented during the First World War. But Chase also provided a more radical, qualitative argument to tackle waste – a change in the economic system from one based on property rights to one based on function. He saw this alternate description of the economy as the strength of his inquiry (39). Quoting English economic historian R.H. Tawney, Chase writes that a functional society, meaning one in which industry's first priority is to address human wants and where any profit-making is a secondary concern, is at odds with what Tawney called an "acquisitive society,"[77] where property is a right and is independent of function in society, and where meeting basic needs becomes a by-product rather than the main goal of economic activity (23). For the remainder of the book, Chase bases his arguments on the "functional conception of industry," and frames solutions to reducing or eliminating waste not primarily as a moral problem, but more so as an engineering challenge (24).

77 Chase is referring to R.H. Tawney's book *The Acquisitive Society* (1920).

Chase's Broad Cultural Interpretation of Waste – Not Solely a Technological Problem

Chase begins *The Tragedy of Waste* by illustrating four different types of waste, "vicious or useless" goods and services, lost man-power, excess man-power, and the waste of natural resources (1–3). Chase then mapped the four types, or "channels," of waste onto his theory of waste, which, at its base, wanted to relegate the production of all non-essentials to a secondary status. According to Chase, primary economic activity should go to the provision of the following nine necessities: food, shelter, clothing, education, recreation, government and community control, the safeguarding of health, religion, and art forms (30). This step in his analysis, to address human wants first, illustrates the radical manner in which he hoped to see the cultural scope of technology expanded to serve society. The other elements of his theory are distilled from the channels of waste listed above into concrete categories: wastes in consumption or what English writer John Ruskin coined "illth," wastes in labor resources, wastes in the technique of production and distribution, and wastes of natural resources. What Chase meant by these phrases is explained in detail in the next few paragraphs. It should become evident from Chase's analysis of waste that he suggested a reordering of relations and values in society. This broad goal went beyond the technocratic idea of using "technology as social solution" (Marcus and Segal 179–206).

Useless goods and services make up the first category of waste. Labelled the main classifications of illth, as laid out by Chase, they include the "military establishment, harmful drugs, narcotics and patent medicines, commercialized vice and crime, adulteration of food, clothing, housing and other necessities, speculation and institutionalized, quackery – sure cures, get-rich-quick performers, super luxuries and their cheap imitation, fashions, commercialized recreation and advertising" (*The Tragedy of Waste* 31). Another aspect of waste in consumption or economic loss mentioned by Chase is when ignorant or superstitious consumers buy so-called necessities. In effect, identification of this form of waste prompted Chase to call for consumer education for the necessities (food, shelter, fuel, clothing) based on the principle of maximum value at minimum cost (32). Of note in this list is Chase's opposition to war and the drug industry, but also his disdain for the bulk of advertising, or, as he refers to it later – the technique of advertising – as the "big brother" of all forms of illth (108). Chase was clearly making the case to limit the use of certain technologies in society. In what could be read as an early critique of (dis)information technologies and its effects, Chase called advertising "artificial stimulation," and warned that it gives a "certain illusion, a certain sense of escape" and creates a "dream world" that does not match the reality of most working people (125).

Waste in labor resources or idle manpower make up the second category of waste. Types of unemployment differentiated by Chase include seasonal unemployment, cyclical unemployment, residual unemployment, turnover losses, strikes and lockouts, time lost by preventable accidents, time lost by preventable sickness, absenteeism, lost labor of the idle rich, lost labor of the unemployed (33). The only issue with regard to technology that eventually emerges from this category is the discourse around technological unemployment; however, it is not part of the discussion in the book.

Waste in the "technique of production and distribution" make up the next category. It should be noted that in the section heading referring to this category, where Chase could have used technology, he uses the term technique, which shows the inconsistent usage of terms at the time. In this third channel described by Chase, he subdivides the main categories of waste into: "co-ordination failures, excess plant capacity, duplication of services, restriction of output, commercial failures or bankruptcies, lack of standardized practices, grades, commodities, bad internal management, lack of cost systems and research facilities, secret processes, suppressed inventions, and wastes in the technique of distribution" (36). Counter to a narrative that promotes technology driving history, Chase infers that the best method of doing something cannot always be equated with the state of the technical arts. He specifically cites the disadvantages and costs of having to constantly replace machines or processes with the latest technical machines, methods, and processes.

Waste of natural resources constitutes the last category. Chase defines destruction of natural resources rather vaguely as "over and above the needs of prudent current consumption," but warns of their use until exhaustion (36). The main resources that Chase lists are: "coal, waterpower, oil, natural gas, mineral ores, timber, soil, animal life including fisheries, and failure to utilize by-products" (37). While he does not trace the historic development of waste of natural resources, Chase's inclusion of this category represents one of his boldest attempts to hold what could be called technologies of consumption to a different cultural standard (236).

A Different Technological Imperative – Living, Not Just Existing

Returning to the notion of the uniqueness of his contribution to early discourses on technology, in a chapter entitled "Industrial Co-ordination," Chase combines his critique of the negatives of technology with his argument for economic planning on a national scale. Coordination, which he defined as "planning for a given end," would involve usurping corporate control of technology (176). As a student of economic history in the mold of Veblen, Chase was aware of many cases in which business interests worked to undermine technological advances (191). Chase emphasized the "restriction of output" or other deliberate measures that businessmen can undertake to control production and distribution in industry (189–190). He summarizes Veblen's line of critique as "the conscientious withdrawal of efficiency" or "capitalistic sabotage on industry." This is when, for example, large corporate interests that have monopolies maximize profit by controlling output. Instead of high-volume and low cost, volume is moderate and costs for consumers remain relatively high (190).

For Chase, technological change involved waste and inefficiency. He lamented the situation in 1920s America: "it almost seems as though there were a relentless law at work which, with every gain in invention, every improvement in technique, threw off a stream of parasites to eat up the slack, and leave us where we were" (275). Offsetting any advantages gained through the technical arts, Chase saw waste as preventing the flow of useful goods to consumers. Yet equally as important, he understood the examination and subsequent elimination of waste as giving people in the United States an opportunity to pursue interests other than consumption to "develop whatever creative impulses lie within" or what he referred to as "the life more abundant – for living instead of existing" (276). To close the book, Chase appeals to engineers, scientists, and statesmen to bring about his vision of a less wasteful and more centrally controlled economy (280). He calls this the challenge of waste.

"Imaginative Regard for the Past" – A Matter-of-Fact Assessment of the Technological Present

A few years later, the challenge of waste had morphed from a description of the problem into a historicization and evaluation of the problem, or what has been called "technological assessment" (Jamison 87).[78] True to his accounting background, Chase created a balance sheet of advantages and disadvantages of the machine age. As Lewis Mumford would later point out, with the exception of Veblen's work, before 1929, the "American literature of technics was singularly sparse" (*Sketches from Life* 449). This is why Chase's next book, *Men and Machines*, which Mumford later in life called "pioneering," is of particular significance. Chase began his "unofficial survey of little explored regions" with a dichotomy between slaves and philosophers:

78 Chase believed that one needed an "imaginative regard for the past" when asking about the purpose of an economic system (*A New Deal* 150–151).

"certain philosophers hold that machinery is enslaving us" (*Men and Machines* 1, 21). Rejecting the first position, he then divides societal responses to the process of mechanization[79] into three philosophical camps, philosophies of "ecstasy," "gloom," and "fence rail" (19).[80] Intrigued by the exponents of all three alternatives, Chase posed questions that search for a better understanding of technology, by which he meant the machine or machine culture, its historical origins and laws, and its structure and order. Yet beyond these theoretical questions, Chase problematized the major social effects of these changes.[81] Not one to fully praise or denounce his subject matter, Chase argued that the best way to assess the situation of men and machines was through the study of the technical arts, or as he phrased it, "the clearest path is in technology . . ." (21).

This now antiquated use of the term technology, the study of the technical arts, is an example of the multiple meanings of technology employed by Chase in *Men and Machines*. There are at least three other different meanings. They are representative of both the ambiguity in the term and the emergence of its usage in Chase's vocabulary, including in its modern meaning.[82] A second meaning was referred to in the paragraph above – a regime or order – as in the sentence, "The level of monotonous work under the machine technology for great classes of occupations is held by Mr. Beard to be not worse but better than in the handicraft

79 Chase defines the process of mechanization at the end of the book as "the billion horsepower, the factory system, mass production, the automatic process, the whole technology so far achieved" (*Men and Machines* 330–331).

80 The three section headings indicating these positions along with their proponents, some of which had very problematic writings on other subjects, incl. race, are: "indictments of the machine" by authors Samuel Butler, Austin Freeman, H.G. Wells (*Men and Machines* 9); "the brave battalion" by authors Henry Ford, Charles Beard (15); "the fence rail" by authors John Dewey, Harry Elmer Barnes, W.F. Ogburn (18).

81 Chase asks probing questions such as: "Are machines prolonging the average life span only to raise the percentage of invalids and neurotics? What are they actually doing to craftmanship, skill, quality of output, art, architecture and recreation? Is social intelligence on the decline? Is social standardization a fact, and if so, is it any worse than standard taboos and mores of other cultures, and of primitive peoples? How much justice is there in the claim that the sublime spiritual values of the East are crucified by the materialism of the West?" (*Men and Machines* 20).

82 It should also be noted that Chase only used the word *technology* once in *The Tragedy of Waste*, and in yet another meaning – the state of the industrial arts. With regard to manufacturers, Chase writes "They have no specific department devoted to analyzing their product, and to keeping up with new invention and technology" (*The Tragedy of Waste* 161). There are at least 19 mentions in *Men and Machines*, which include not only the three other meanings detailed above, but also its modern usage. The difference between the publishing dates of the two books – 1925 and 1929 – clearly indicates the emergence of the term *technology* in Chase's vocabulary to describe changes in society.

eras" (172). A third meaning indicates an era like in the sentence, "In the last analysis more economic independence is to be secured in a machine technology, humanly controlled, than ever obtained in the handicraft era" (330). A fourth way in which Chase uses the term technology corresponds to its modern meaning like in the sentences, "It marked the beginning of the great period of Egyptian technology" (45) and "Technology is developing at an incredible pace" (292).

Chase's study of technology, to use the term in its current meaning, opens with fundamental questions about the "anatomy of machinery" (22). In the second chapter of his book, Chase discussed the distinction between a tool and a machine and ended up using the odd construction "tool-machine." Citing anthropology, he provided a broad definition of machine to be "any non-living contrivance to extend or modify the power of the body, or to refine the perception of the senses" (24). His concern, despite overly deterministic prose,[83] was not with machines, but with the effects of exponentially increased power. The origin of this advance in power he dated to the invention of the steam engine in 1775, but he also credited the adoption of Arabic numerals and cipher, as well as to the introduction of scientific methods. From these short descriptions it becomes rather obvious that Chase's intended audience was not an academic one.

"Superficial but Suggestive"

Lewis Mumford's characterization of *Men and Machines* as "superficial" in the bibliography of *Technics and Civilization* accurately describes Chase's brief historical account of technology.[84] In only three chapters, he surveys 7,000 years of human ingenuity. Beginning with the machines of the ancients, it amounts to a few practical applications of knowledge during Egyptian, Greek, and Roman times. While significant as one of the first attempts by a U.S.-American author to historicize the social effects of machines, it largely tallies the positive and negative aspects of machines and machine culture through narratives of inventors and inventions, techniques, and tools.

83 "The concern of the Machine Age is [. . .] with those particular sorts of tool-machines which are doing new and unprecedented things to modern civilization; things which no earlier civilization ever experienced; which are changing timeless concepts and habits, engendering unheard-of behavior patterns" (*Men and Machines* 24).
84 This is Mumford's characterization of chapters four, five, and six of *Men and Machines* (*Technics and Civilization* 453).

Despite its lack of sustained historical analysis, Mumford also found *Men and Machines* "suggestive" enough to pursue one such question that Chase did raise: why the so-called industrial revolution had to wait until eighteenth-century England to occur. It was a question that Mumford would later expand on in more depth. Chase, in his initial answer to the question, indicated that many early civilizations had slave power to do most of the work. He even goes so far as to suggest an inverted relation between technological development and number of enslaved people in these early societies. Chase notes that in Greek society, attitudes toward what was then known as the vulgar arts were mainly negative. This is mainly due to the intense labor required to do all things technical. Chase quotes Xenophon of Athens to demonstrate the perceived lack of will to apply oneself to or develop the vulgar arts in Greek society:

> the arts which men call vulgar are commonly decried with good reason. They utterly ruin the bodies of workers and managers alike, compelling men to lead sedentary lives, and huddle indoors, or spend the day before the fire. Then as men's bodies become enervated, so their souls grow sicklier. And these vulgar crafts involve complete absence of leisure, and hinder men from social and civic life (*Men and Machines* 50).

A second answer Chase provided as to the beginning of the industrial revolution follows in the next section of his book. Approaching the Middle Ages, Chase recounts the beginnings of science starting when the Moors pushed into Spain in the ninth century, which brought a "fresh breath of new knowledge to the medieval world" (*Men and Machines* 57). He later asserts that "their Arabic numerals and precision instruments" became the "first sparks" of the industrial revolution (67). He then turns to emphasize the importance of Roger Bacon and Leonardo da Vinci as well as the significance of time measurement (the clock) and its massive effect on European society. Chase briefly mentions the large improvements to machine-tools during the Renaissance, yet beyond this traditional account of the development of technology in history, he includes a method/economic model as part of his historical account of technology. He refers to it as the "Venetian approach," or the water-based trading and transportation of manufactured goods between East and West (63). Rather than belittling the once vulgar arts, the trading approach embraced them to gain economic power. Important to his overarching storyline, according to the author, was the need for a different kind of power. Wind and water were soon to be replaced by another prime-mover, one which would power the manufacture of goods and increase the speed of their transportation.

Chase's brief synopsis of the Industrial Revolution spans two chapters, starting from James Watt and ending at the mechanical man. On these pages he outlined how power was introduced to the cotton, coal, and transportation industries. He indicated that, in order to have a more detailed account of the industrial revolution, he would need to include many other industries like "pottery, printing and paper and lumber" (76). Yet as he pointed out, the history chapters in *Men and Machines* were not intended to compete with more rigorous historical accounts. They were "not a history but only a layman's glance at Watt and his friends; a glance before the avalanche of new inventions descends in a matted mass requiring the most exhaustive research to untangle. The broad outlines of the revolution have been described by many competent historians, and to them the curious reader is referred." With expectations lowered, he proceeded to provide a third explanation as to the causes of the industrial revolution in the form of a rather thorough list of contingent historical and environmental factors (68–69). Implicit in this list is a complex understanding of historical causality. For Chase, technological change is principally as result of socio-cultural forces. First, he demonstrates an understanding of wealth, power, and individuals' ability to justify certain systems and courses of action. He attributes much of the more recent drive to develop technology not to a certain deterministic law or universal improvement of man, but to individuals with a profit motive as is evident in the following passage:

> Engines did not come to save men the brutality of dull labour, to help an Emperor build an imperial city, to irrigate and replenish waste lands—they came because English traders wanted to increase their profits by making cheaper cotton to sell abroad. The human values, such as they were, had to enter by the back door, and despite much lofty talk to the contrary, this is the route which they still follow (69).

Another example of Chase's contingent framework of historical understanding is when he describes invention as a process by which many past ideas are incorporated into a single new invention. He also self-critically ascribes less importance to the necessity of specific individual inventors or personalities than many authors of his time might have ventured to do:

> No more effective illustration could well be given of the fact that invention is normally a social process, rather than the work of a few great men. We writers try to make our pages dramatic by concentrating on the geniuses, but inevitably we distort the story. Even Watt might have spent his life on chronometers and organs, had it not been for Newcomen; while if Watt had died at birth, the chances are that a condenser would have been invented before 1800 (73).

After describing the emergence of industry in various national settings, Chase closes his historical account detailing mass production, automation, and standardization of designs, processes, and parts. Despite his characterization of his book as non-historical, Chase had written one of the first, simple but nevertheless suggestive histories of technology using a contingent framework to describe technological change.

Framing the Impacts of Technological Change as a Disequilibrium

At the beginning of chapter six, Chase pivots from a general historical summary to state the central theme of the book: "our main concern . . . is not with [machines'] technology[85] or history, but with their impact upon the day-to-day life of human beings" (108). In the process, he switches from recounting a distant past of machine-tools to a concern with the "present past"[86] of labor history in the industrial era. This move creates a bridge of continuity in the book from the past to the present. With regard to historical causality, Chase understands the "impact" of machines and machinery as being both direct and indirect; that individuals are "influenced by" direct contact with machines, but more significantly, that society is influenced by the "new commodities, services, laws, folkways, standards, philosophies, arts" that this cultural environment "has engendered" (108). The most direct impact of the machine,[87] Chase writes, is in the habits of work, play, and environmental and noise pollution. Beyond immediate sensual experience, these individual factors interplay with social factors to create social systems. An example of such is the factory system in eighteenth-century England. Chase was concerned with the effects of the factory system had on individuals' lives and freedom:

> The factory system is . . . a legal system, whereby the building, the raw materials, and the tools are owned by the factor, or capitalist; the worker contributing nothing but his labour. This normally makes for more efficient production than cottage industry, but few have claimed that it made for greater happiness or independence. The time-clock displaces one's own sweet will (126).

He continues this line of argument in another passage questioning the balance of life after the advent of industrial structures:

> Nobody could argue that a peasant or a journeyman before 1750 was a freeman, but he was infinitely more free, and infinitely more of a human being, than his grandchildren in the mills and mines of 1830. Life was more versatile and interesting when craftsmanship was combined with agriculture. For many occupations in the new system only one set of muscles and one part of the mind was used, and that a low one; other faculties remained idle and undeveloped. Nor were there any schools or playgrounds to right the outraged balance. The Greeks and Romans built immortal public / buildings and made provision for public amusements. Manchester was very different (129).

85 Chase's usage of the word *technology*, in this instance, refers to the study of technics.
86 This term is explained in appendix 2 at the end of this book.
87 When Chase refers to "the machine," he is using a synecdoche for machine culture.

While Chase ignores the brutality of the enslaved peoples and class system that made building in ancient cultures possible, his reflection on the lack of balance in nineteenth century England points toward a larger argument. Summarizing the mainly negative outcomes during the era, he wrote, the "gain in power was turned into a social loss"(131). As a critical historian-author, Chase questions not only the quality of life under the factory system, but also the human role in technological change when writing about the population of Lancashire, England. In the passage below, he rejects the notion that any type of *theos apo mechanes,*[88] or god from the machine, would somehow swoop down and interfere to correct a problem beyond human solutions or rearrange technological change for the betterment of the general social welfare:

> There was no conscious attempt to adjust the worker to the exigencies of his new task; the town to the pressure of its new population. Properly guided, the early advantage might have been held, but the very forces which made the industrial revolution possible, expressly stipulated that they would brook no interference from God, nor state, nor man, in the unhindered pursuit of their personal advantage. Had not Adam Smith, himself, assured them that an "invisible hand" reached forth and so arranged terrestrial affairs that the minimum of social interference automatically insured the maximum of social good? (127).

To combat the "disequilibrium" that was created during the era, Chase mentions major reforms that were fought for and implemented after 1850 to improve working conditions, namely factory legislation, civil service (factory inspection), and trade unions (134). In addition to these counterbalancing developments, Chase adds major increases in population, the spread of specialization, and the economy (and importance) of cash as other broad outcomes of "steam culture" (137). Despite these the developments, imbalances persisted. One last, but major result of changes in technology mentioned is what Chase refers to as the "most important change in social habit," namely a reduction in "economic security and independence" (140).

The Loss of Cultural Balance

Throughout the next few chapters Chase shows a preoccupation with balance within a framework of dynamic technological change. This is evident in his analysis of the physical and mental health of workers, the degree of satisfaction at

88 Literally, "god from the machine," which refers to a human actor of a drama, dressed up as a god and hanging from a crane, who would resolve a problem that was beyond human capacity (*Men and Machines* 50).

work, the imbalance between saving labor in production but not in consumption, the notion of technological unemployment, a better balance between staples and new commodities, and the quality of goods and information flowing to consumers. This theme of balance is also present when Chase switches to the topics of art and play in the subsequent chapters.

In a chapter on robots, which are defined as machine tenderers, Chase looks at types of machine workers, at the health of the industrial worker and at industrial accidents. As he notes, the initial effect of the machine was to hurt the worker physically and mentally (166). Despite health improvements within the past generation, Chase thought that the machine's effect on the workers mind, a mental imbalance, had been prolonged. He introduces the notion of the philosophy of fatigue in his account and acknowledges the appearance of "unwilling robots," people who do not adjust to modern work (167). In addition to suggesting automation to replace toil and monotony, Chase recognizes that even from an economic perspective, it pays to safeguard the health of the worker. Part of that idea was also acknowledging the effects of an increase in specialization in industry. While he could find no evidence of a reduction in skilled workers, just a switch to new modern skills, Chase confirmed that specialized work was neither an advance nor a retreat in culture but did appear to be coupled with less work satisfaction. It is at this point that Chase cites the paradox of machinery. Instead of saving labor, more machines had meant more labor, new necessities, a complex new establishment in which to operate, a highly intricate infrastructure in modern cities, and other labor-intensive novelties like shipping raw materials over great distances.

In challenging the goals of technological change, Chase remained true to the template that Veblen had provided. His criticisms were not necessarily of technology itself, but of the "going economic structure" under which technology was developed (193). Chase believed that modern industry was saving labor in production costs, but not as much in distribution costs (201). In short, "more stuff, more congestion, . . . more cross-hauling" may have meant progress to some, but it certainly also meant more complex and difficult work for many (203). The theme of imbalance was extended to jobs. Chase seriously thought that machinery was displacing human workers, an issue known as "technological unemployment." He also believed that it would increasing become a problem. The solutions he proposed, from a reduction of working hours, to increases in production of necessities and comforts to abate poverty and hard work, were being ignored, he wrote. Instead, in line with a Veblenian critique of industry, Chase put the blame on the "present financial control" that was not allowing enough purchasing power (low wages) and not promoting the quality of necessary goods and services that would make for a good life. Chase indicated that an inverse relationship between increasing mechanization and decreasing purchasing power would mean that improved production would result in being worse off. He called this unbalanced arrangement the "economy of a madhouse" (216).

To characterize his stance against the flood of manufactured goods in the 1920s, Chase quotes the Greek philosopher Diogenes: "a man's wealth may be computed in terms of the number of things he can do without" (218). Chase argued for a better balance between staples and new commodities, for better quality of goods instead of willful obsolescence or deliberately encouraging waste in consumption (230). The industry policy of appealing to style first, price second, and quality last meant for Chase that the "technological structure" had to be reformed and "machine process" had to be more intelligently controlled (238–239).

As his next chapter title – "Skyscrapers" – suggests, like many other observers at the time, Chase believed so-called advances in technology were bringing about new forms and a general renewal in the arts. Such a rebirth of art in a machine aesthetic, Chase believed, should be used to counterbalance the complexities of living in the modern age: "technology demands not more complications, but simplification in the form of parks, open spaces, radial boulevards, perspective, sun, air, light. In such an environment, the machine can really function, and give us an architecture of such towering loveliness as to make the heart falter in its beat" (247). With regard to the negative effects of machine culture on art, Chase admitted that there was a greater gap between work and art and therefore that many workers with over-specialized activities were cut off from any creative function (241). For Chase, a counterbalance could be found in re-creation or creating again in play what has been lost in work. He contrasted this notion of play with "decreation" or diversions that compounded the lost balance through unrewarding forms of play.

Despite the positive effects of machine culture on society, Chase remained pessimistic about the loss of cultural balance for the individual. As he summarized, Chase thought that work in a mechanical environment created a psychological imbalance:

> In our analysis of robots, skills, the flood of goods, and even art, we found that the gloomy prophets, never troubling to look below the surface, had tended grossly to overstate their case. There was much that was depressing, but the hopeful elements were also many. In play, however, we have struck upon a vein of pessimism that is difficult to disperse. Americans are the chief proponents of the machine, and they are not a happy people. The fact that they are restless, nervous, at odds with their world, and frequently depressed, does not need to be argued. Look at the faces in the streets of any city. Play is the flywheel of life, and America, with the most stupendous recreational equipment ever dreamed of, does not know how to play. It can only step on foot throttles, insert coins in metal slots, scan head-

> lines, crowd through clicking turnstiles, rush head long down roller coasters—seeking a balance which these things can seldom give. This is the more tragic, in that the specialization and frequent monotony of our daily work, however skilled, clamour for psychic equilibrium (263–264).

As this quote clearly illustrates, Chase understood the negative psychological toll, the stress, of living and working in a mechanical environment. He also recognized the importance of a certain kind of play, as opposed to fun or excitement, to reduce these effects.[89] Because machine culture had destroyed many habits and standards that had been built up over time, Chase thought that not only individuals, but American society as a whole had lost its "cultural balance" (272). The situation called for adaptation and readjustment to dynamic changes, with three exceptions.

Technological Tenuousness

The three great potential dangers, or "leaning towers," of machine culture that Chase did not think that American culture could deal with were the dangers of mechanized warfare, the growing complexity and danger of mechanical specialization, and the mounting drain upon natural resources. First, Chase warned against the "two-hour war" in which an aggressor could quickly annihilate an opponent. He was not only against the use of technology in war, but also advocated for a ban on war altogether. Second, Chase alerted his readers to the tenuousness of the technological structure. He called this phenomenon "technological tenuousness" and warned against a dramatic event, whether by humans or by nature, that would cause many systems to collapse. Chase wrote that the connection with the realities of the technological environment were tenuous, like in the example of the one-industry town. Once an industry decides to move out, the people in that town are left to fend for themselves. Beyond such sudden changes, Chase also cautioned about the complexity of various technological systems that no one person could understand in a discerning manner. Finally, Chase wrote that the exploitation of natural resources like oil, natural gas, coal, iron, copper, soils, lumber, and fisheries was not sustainable.

89 Chase had written a chapter in which he examines the way in which the machine age effected new and old forms of play in *Whither Mankind A Panorama of Modern Civilization* (1928). As the title of the book suggests, Chase observed many negative elements of play in 1928 America. The collection of essays was edited by Charles Beard and included among its many pages and authors a chapter on science by Bertrand Russell, a chapter on philosophy by John Dewey, and a chapter on the arts by Lewis Mumford.

If the technological structure of society could be characterized through its instability, Chase called for an assessment of the relationship between technology and society. In a chapter entitled "The balance sheet," Chase groups the effects of the machine into those good, those evil, and those that have a combination of both. The noticeably good effects were an increasing life span, higher living standards, shrinkage of space, a decrease in class distinction based on land ownership and patents of nobility, a decrease in hours of labor, a decline in superstition, the expansion of the ego gained, for example, through driving, introduction of a philosophy of fatigue to deal with workplace hazards, an increase in exchange between pure and applied science, and an increase in social sympathy and a decrease in cruelty (319–321). As decidedly evil effects Chase listed mechanized warfare, the tenuousness of connection and balance in the interlocking industrial structure, the exploitation of natural resources, monotony, specialized tasks that upset the work-play-art equation, the importance of money (symbols replace reality), workman displaced by machinery, competitive imperialism, the fact that machinery has created a new ruling class, an increase in mental illness, an increase in accidents, an overvaluation of industry, city congestion, noise, dust and smoke, the disastrous first effects of introducing machine culture, and the degenerative impact on indigenous peoples (322–325). The effects both good and evil were migration from farm to city, the loss of self-sufficiency, the uprooting of old skills and the creation of new ones, the entrance of women into the wage-earning class, women and children in the factory and classes of people gaining useful work, mass production that allows for higher incomes for youth, the break-up of mores in power age, uniformity with frequent and drastic change, changes in art, the increase and decrease in quality of goods, an increase in written documents and literacy, a faster pace of life, the loss of economic independence but increase in standard of living (325–331).

On the whole, Chase believed that machinery (what would now be called technology) had brought "more misery that happiness into the world," a view seemingly at odds with progressive doctrine. He observed in technology both benefits and detriments. But his overriding sense was that, as currently utilized, technology was both tenuous and dangerous. If, however, usage could be modified to maximize technology's beneficial potential, then the advantages would far outweigh the dangers.

The Statistical Don Quixote Declares His Independence

What can be said about Chase's conception of technology and his observations about a changing world in *Men and Machines*? In the book, Chase characterizes himself as a guide along a quixotic quest.[90] Much as a comedy sets two opposing forces in opposition to one another in a struggle, Chase attempted to juxtapose older, pre-industrial forms of living and working with new forms of machine culture and the power age. The only part he leaves out of his account is the final resolution of the story, which is to be determined by future generations' ability to control the development of technology. Instead, Chase furnishes the last chapter of the book as a dystopian thought experiment in which he wrote three alternative endings: drifting with the "tide at present," adoption of a "simple formula like government by business or state socialism," or facing the "full implications of the machine" with intelligence and by possible creating a new formula to conquer and control new challenges and new machines and powers as they appear, or as Chase calls them the "billion wild horses" that have sprung from the human mind (347–348). The first ending probably refers to the continuation of an economic structure largely at the mercy of the ebb and flow of machine culture laid out in the book. Chase thought it was also the most likely. The second, Chase already saw as failed, providing the example of the switch from a Marxian credo as part of the Russian Revolution to the New Economic Policy that followed. The third Chase envisioned as a "boldest, most exhilarating, most dangerous adventure that ever challenged the intelligence and spirit of mankind" (348). He wanted society to face the consequences of dynamic technological change.

In his notes, Chase later characterized the final two pages of *Men and Machines* as ending on a "pragmatic planning note,"[91] but as Taylor society[92] member Walter Polakov explains, this approach (pragmatic planning) was unnecessarily confronting a reality that no longer existed. Much like Cervantes' Don Quixote, Chase's self-

---

90 Self-referential characterization of Chase as a guide along a quixotic (idealistic) quest (*Men and Machines* 21).

91 On a notecard, Chase acknowledges his own misunderstanding on the last pages of *Men and Machines* with regard to money and credit (pages 344–345) as being too functional in nature. The notecard ends with the statement, "I end Men and Machines on [a] pragmatic planning note" (The Papers of Stuart Chase. Manuscript Division. Box 15, Notes. Library of Congress, Washington D.C. 3 January 2020).

92 According to their bulletin of the Taylor Society, their goal is to promote the "science of management."

reference makes him vulnerable to the criticism that he may not have fully understood the fundamental changes to conceptions of technology and society in his time:

> Don Quixote fought windmills, taking them for vicious giants; he adored the old scullery maid Dulcinea, fancying her his inspired genius. He was a symbol of the dying age of knighthood – his distorted mind, unable to grasp the social-economic changes of his time, made himself a figure grotesque but likably pitiful" (Polakov 19).

Chase references the era known as the Power Age in *Men and Machines*,[93] but Polakov argued for further acknowledgment of this concept in his 1933 monograph entitled *The Power Age*. This new age could be characterized not by "bigger and better machines than those of the Machine Age, but by a different kind of machine, conditioning a different kind of production, which results in a different kind of economy, and a different kind of social relations" (Polakov 18). This line of argument led Polakov to question whether a seemingly quixotic approach to archaic notions of technology was the best way forward: "Can that much be said of the modern knights of the dead Machine Age, who unwittingly or otherwise advocate a New Deal with the same old pack of marked cards?" (19). "A world without slave and lords," meaning with more economic freedom, according to Polakov, could only be achieved through a better understanding of the modern world. As he wrote, it will "remain a vision of the few and not a reality for all, as long as we continue to live among all our creations of science and engineering as maggots in a cheese, knowing nothing of the nature of the world we live in" (21).

Polakov's criticism could be understood as a swipe at Chase, but it was he who had taken a first step to achieve a new relation between capital and technology in society, first in his 1931 article "Declaration of Independence" and subsequently in his 1932 book *A New Deal*. In the 1931 article, Chase invoked the Veblen template to declare his independence from "the dominant ideals and philosophy of the business community." Having framed the technological structure of society as tenuous and subject to contingent change in *Men and Machines*, Chase felt justified in choosing to shape his own relation to technology, one of action on quality of life improvements, and to reject the "fragile, impermanent" goal of "acquisition" ("Declaration of Independence" 29). His declaration of independence was two-fold. First, he desired independence from the "commercial and material formula of wealth and happiness" (30). Chase had recognized a large gap between the sophistication of material things, which was far in advance of the sophistica-

93 See page 25 of *Men and Machines*, where Chase comments on a quote from Henry Ford about the origins of the Power Age.

tion of an American philosophy of life that would place emphasis on life sustaining elements like proper recreation (31). Second, he wanted independence from "self-depreciation" in the form of leisure that "complicated existence" and the "imported aesthetics" of conspicuous consumption (30).

Ultimately, Chase was arguing for a change in the way economic activity was oriented and structured. Attempting to "sharpen the line" which Thorstein Veblen had drawn between industry and business, the main point of Chase's declaration of independence was to salvage the conception of technology from vested interests: "from now on," he wrote, "industry must become an art" (36). As elaborated in *A New Deal*, the scope of that art was certainly a national one. In his personal notes, Chase wrote that, "Technology demands [a] national economy rather than [an] individual [one]."[94] The historical record shows that Chase was right, at least according to the popular vote of the U.S. presidential election shortly after *A New Deal* was published. Judging by the appropriation of the book into the Democratic platform in 1932, the basic question posed in the book's opening chapter – "What is an economic system for?" – aptly captured the Zeitgeist under the Roosevelt Administration. Advancements in technology had undoubtedly revolutionized production, the questions that interested Chase and many others at the time was when technology would be used for equal distribution of necessities like well-paying jobs, health care and education as well. As for the origin of the phrase, Chase claims not to have coined it:

> The facts are these: in February 1932 I completed the manuscript of a book to be published by Macmillan to which I gave the title "A NEW DEAL." I had some preliminary publicity of course and in May and June of 1932, the New Republic ran five articles taken from the book under the caption, A NEW DEAL. In July 1932, Franklin Roosevelt flew to Chicago and on the floor of the convention first used the phrase. In his acceptance speech were one or two sentences which had a faintly familiar ring leading me to wonder out of idle curiosity whether those who had helped him on the speech had been reading the New Republic.[95]

In another letter, he denies paternity of the phrase, but acknowledges his contribution to the general approach enacted under the New Deal:

94 The Papers of Stuart Chase. Manuscript Division. Box 15, Notes. Library of Congress, Washington D. C. 3 January 2020.

95 Chase, Stuart. Letter to Mildred Schwarz on 16 March 1939. The Papers of Stuart Chase. Box 1, Correspondence, Miscellaneous 1924–1955. Library of Congress, Washington D.C. 3 January 2020.

> The phrase is a common one and ancient. I think you would be wise not to credit me with any part of it. So far as the content of the book goes, I'm glad to realize that the Middle Road approach[96] which I recommended so strongly was later adopted by the [FDR] administration. But here again many others contributed to the philosophy.[97]

Chase's Legacy – An Ecological-Economic Analysis of Human Plight in an Industrial Society

It may be tempting to dismiss Chase as a popularizer of complex issues, but the Stuart Chase Papers at the Library of Congress tell a different story. This was a man who, for whatever he lacked in dense and novel academic theory, rejected many business opportunities to write about issues that were important to him. Chase's papers, notes and correspondence not only reveal the voraciousness with which he read and took notes, but also how "high up" his influence extended. Correspondence with politicians, most notably Franklin Delano and Eleanor Roosevelt, are evidence of his influence on public policy, particularly from 1932 onwards and after World War two. While it is beyond the scope of the current chapter to discuss Chase's contribution to the political philosophies and economic policy-making activities during the New Deal Era, it should be evident that he should be credited for making the jump from theory to practice.

Perhaps Chase's most prized legacy with regard to the history of technology can be linked with his impact on one of his famous interlocutors. In a letter dated 9 September 1971, Lewis Mumford wrote that he owed a great debt to Chase.[98] Chase had alerted Mumford to a factor in economic thinking that had elsewhere been called "habitability." It is the relationship between gross national product and quality of life. Mumford not only considered Chase a partner in "hammering on this theme for half a century," but praised his *Tragedy of Waste* as a "classic statement with few rivals," Later in the letter, Mumford admitted that he had always felt "an underlying tie" with Chase, even in times when they had very differ-

96 In *A New Deal*, Chase described three approaches to revise the economic structure of the United States. The first option to change the economic system entailed a violent communist revolution. The second involved a big business or fascist dictatorship. The so-called "third road" endeavored to drastically, progressively, and collectively control the economy from the top through a series of major agencies; all while avoiding a total break with the past (235–241).

97 Chase, Stuart. Letter to Ruth Woodruff on 2 March 1954. The Papers of Stuart Chase. Box 1, Correspondence, Miscellaneous 1924–1955. Library of Congress, Washington D.C. 3 January 2020.

98 Correspondence with Lewis Mumford. The Papers of Stuart Chase. Box 11, Subject File Correspondence 1970–77. Library of Congress, Washington D.C. 3 January 2020.

ent opinions. A few years after that correspondence, while going through his files to complete an autobiography, Mumford was reminded of the importance of Chase's contribution to thought in the 1920s. In a letter written to Chase dated 10 October 1973, Mumford praised *The Tragedy of Waste* as the "first ecological-economic analysis of [human] plight" in an industrial society.[99]

99 Correspondence with Lewis Mumford. The Papers of Stuart Chase. Box 11, Subject File Correspondence 1970–77. Library of Congress, Washington D.C. 3 January 2020.

Chapter 4
Converting Technology into Human Culture
Lewis Mumford's Social – Aesthetic – Moral Relations to the Past

If history is more than past politics it is also more than past technology.
Lewis Mumford, "History: Neglected Clue to Technological Change"

Man lives in history; he lives through history; and in a certain sense, he lives for history, since no small part of his activities goes toward preparation for an undisclosed future.
Lewis Mumford, *The Pentagon of Power*

Emphasizing the Technodrama in History

Lewis Mumford was a prodigious scholar and cultural critic whose life spanned almost a century. Born in 1895 in Queens, New York City, in 1936, Mumford moved to the town of Amenia, New York, where he resided until his death in 1990. Mumford wrote on a number of topics, from city planning to architecture, from literature to art, but he is widely known for his 1961 monograph *The City in History*, which helped establish the city as a subject of scholarly inquiry. Aside from contributions to the historiography of architecture and urban spaces, his writings on technology and culture are foundational for the academic disciplines of American Studies and the History of Technology.

Foremost an author-historian, Mumford observed and studied large-scale mechanical changes during his lifetime. He was certainly well-read, but he also refined his own ideas about technology and society through daily writing sessions. Both of these practices ultimately led him to question technology's supposed role in shaping history (understood here as reality). As the editor of *The Lewis Mumford Reader*, Donald L. Miller, puts it, Mumford attempted nothing less than to interpret "the century of science, war and the machine" (5). He held this interest throughout numerous books and articles, despite them having different focuses or topics. On a more personal level, Mumford was motivated to explore the question of how "the world of his youth became the world we now live in." Reading across many of his works one is struck both by the consistency of both his style and of the relations that governed his approach to historiography from the beginning of his writings to the end.

https://doi.org/10.1515/9783111396477-005

When Mumford's *Technics and Civilization* first appeared in 1934, the concept of technology had just begun to emerge as a key word.[100] Mumford thus helped set the parameters of how one could think about technology. As many scholars do, Mumford modified and refined his arguments in later works, which culminated in a two-volume series entitled *The Myth of the Machine*. By his final books on the subject, Mumford had advanced an integrative theory of technology in history. Searching for continuity in formative processes, he colligates evidence across disciplines and from diverse sources, extending and consolidating his earlier conclusions.[101] Over his writing career, he authored numerous articles and books which argue against "civilized history," which he thought overemphasized the political, the economic, and the past, and underestimated the cultural as an explanation for changes in society. In particular, his version of history also aimed at deemphasizing the prevailing interpretation of technology and cautioning against its dominant role in society.[102] Mumford's various accounts contest the technology-as-a-tool metaphor, where the concept is understood as an external object or end product. Instead, technology is conceived as a broad and dynamic man-made enculturation process. This approach means replacing a focus on inventions and objects and placing humans at center stage of historical development. Mumford favored a "theater" metaphor,[103] which broadly views technology as continually being modified by human actors "at every stage of its development by dreams, wishes, impulses, religious motives that spring directly . . . from the recesses of man's unconsciousness" (*The Pentagon of Power* 415). In his mind, understanding the broad scenario and setting of the "performance," Mumford's theory of history emphasizes the "technodrama" to make sense of the all-enveloping process of history. He underscores technology's role as a "formative part of human culture as a whole" and seeks to disavow the *Myth of the Machine* – that

100 Cf. Schatzberg and Oldenziel, as well as chapter one of this book.

101 In the epilogue to *The Pentagon of Power*, Mumford reveals the broad *teleos* of his theory of history: "The Advancement of Life" (414–435).

102 In a 1935 essay which appeared in *The New Republic* magazine, Mumford suggests one such possibility: the subjugation of technology to all of society through communism. "Whereas the rise of Christianity meant the defeat of Roman power, the rise of communism constitutes a vital fulfilment of Western civilization, and marks the passage from a crude faith in mechanical toys as objects of worship to a belief in machines as subordinate instruments of social aims and rational purposes. That faith produces fellowship: this fellowship nourishes action: and action gives a value to the future equal to that which more static and less confident societies assigned solely to the past." ("Civilized History" 63–64).

103 "Man's existence in all its dimensions is perhaps best understood in terms of a theater, as a drama unfolding in action" (Mumford, *The Pentagon of Power* 414).

improving tools to "command nature," thus increasing both power and economic growth, should be the utmost goal of human existence.

Given the scale and scope of his broad inquiry into technology and society, it is important first to navigate through Mumford's influences and formative ideas. This will help readers glean an understanding of how he considered technological change to operate. The historiographic approach employed by Mumford was later classified by Leo Marx as being "organicist" and by Rosalind Williams as "moralist" and is integral to an understanding of how Mumford thought about technology.[104] This chapter expands on such previous classifications through a close reading of *Technics and Civilization*. In this chapter, Mumford is shown to employ modes of historical explanation that are also contextualist and radical. And beyond his moral sequencing of the history of technology, the current analysis establishes how Mumford engages with the past in various ways. Importantly, these multiple relations to the past – social, aesthetic, moral – help reveal Mumford's dialectical position with regard to how technological change operates. He sees human culture and purpose interact with an increasingly mechanical mindset and environment.

In publishing this first attempt at a general history of the machine in 1934, *Technics and Civilization* can be considered part of a network of authors who argued for a more contingent understanding of technological change. On this view, *Technics and Civilization* can be considered a practical (moral) guide to recover a usable past. It was Mumford's hope that, by preferring a past that discerns what he later called "authoritarian technics" from more "democratic technics," his readers would be in a better position to shape processes and practices and culture from within, as opposed to being the "fated victim of external forces and external institutions," which are not fully under their control (*The Pentagon of Power* 421).[105]

Influences and "Confirmation of Intuitions"

Lewis Mumford attended New York's Stuyvesant High School starting in 1909, which at the time was an all-boys "industrial and scientific" school "dedicated to . . . the mechanical arts" (*Sketches from Life* 98–99). From an early age, he was on track to become an engineer, learning the language, skills, and outlook disciplined

104 See Leo Marx, "Lewis Mumford: Prophet of Organicism" 164–180, and Rosalind Williams, "Lewis Mumford as a Historian of Technology in *Technics and Civilization*" 44.

105 Technology was later described by Mumford as part of human expression and an endeavor of "mind-making, self-mastering, and self-designing" (*The Myth of the Machine* 9).

by a background in the hard sciences of the time. Yet, Mumford described his interests in technology as having been "cerebral and imaginative, rather than technologically inventive and practical" (449). After 1912, he had taken the decision to become a writer (Miller, *Lewis Mumford: A Life* xix). In an autobiography of his early years, Mumford still credited the "kind of elementary scientific and technical training" he had received in high school, without which he admitted he could not have written his two major works on the subject (*Sketches from Life* 101).

Once he started reflecting and writing professionally about the changes wrought by the so-called Machine Age, his perspective became both broad and historical. Instead of narrating technology in history in the mechanistic mode, in early 1924 he and others in the fields of history and sociology expressed thoughts on the subject in response to a "distinct drift toward history" (Novak, *The Correspondence* 189). As he revealed in a letter to his mentor, Patrick Geddes, Mumford intended to demonstrate the use of a unique methodology and outlook to analyze the American cultural past, which in the United States in the 1930s, was called a history of civilization: "if we attracted the attention of people to our outlook and methods by demonstrating their use in analyzing the Making of the Past, they might be a little more willing, perhaps, to see the place of our sociology in the Making of the Future!" (189).[106] Seven years later, in September 1931, Mumford wrote that he had a semi-finished draft of a book that was a "sort of conspectus of modern civilization and its possibilities – social, esthetic, moral – beginning with the part played by the machine and carrying the integration upward through buildings, cities, regions, and civilization as a whole" (330). So how did Mumford start from forming ideas toward a history of civilization in the United States and end up with a seminal classic on technology? What was the new methodology that he planned to apply to his studies, what were its origins, and what kind of philosophy of history did Mumford employ, as he said, while attempting to "make" the past?

As a young man, Mumford read works by many scholars he credited as being major contributors to his education, but two stand out, namely Patrick Geddes and

106 Approaching the topic of technology from a larger scope, in this context human development, lends Mumford's writings a non-discriminatory character. As Shuxue Li astutely reminds his readers, Mumford is not exclusive in his analysis of technology and use of terms like culture and civilization. Deemed a "writer-critic" (Li 24), Mumford is described foremost as a writer of wide-ranging, long monographies not bound to one subject area. In his writings on the subject, Mumford does not sketch the historiography of technology as a "promotion of Eurocentrism and white cultural supremacy," but rather as a "critique of Western hegemonic culture" (210). It is this inclusive enframing of the technological landscape that prompts Li to laud the "strong moral and social responsibilities" Mumford exhibits in his works (24).

Thorstein Veblen. A "disciple" and enthusiastic reader of the Scottish evolutionary biologist, Mumford's scholarly interests first formed by reading Geddes's works starting in the fall of 1914, after he had enrolled in the City College of New York. Subsequently, before the two men began what was to become intermittent written correspondence, Mumford also initiated a meeting with the man he would later address as "master" in 1917.[107] In terms of a general literary approach, two broad insights can be observed that stem from Patrick Geddes's influence. The first is the view that scholars should couple their intellectual endeavors with activism, a point that will be addressed later in this chapter, and the second, that a scholar can aspire to be a generalist[108] that could help "bring all the diverse parts of [his] education and [his] environment together and transform them into an increasingly intelligible and workable [. . .] whole" (Novak, *The Correspondence* 24). Intellectually, Mumford gleaned from Geddes's acquaintance a "confirmation of [his own] intuitions,"[109] a framework for his later writings on technology that was informed by biological and sociological principles.[110] An emphasis on biology stressed upon Mumford a reverence for organic life in all its forms, but it also drove him to "overcome [. . .] the forces that threaten the organism" (34). An emphasis on sociology was reinforced not only in conversations and walks with Geddes and his associate Victor Branford, but also in his brief role as editor of the *Sociological Review* from April to October 1920, and in his dealings with the Sociological Society (Mumford, "Victor Branford" 43). Reading about Geddes's activities of city planning stressed the interrelations among diverse organisms and factions into a complete whole.

---

107 Although their personal meetings "got off on the wrong foot; and . . . never managed to fall into step afterward," Mumford and Geddes corresponded off and on from 1917 to 1932, the year of Geddes's death (Novak, *The Correspondence* 345). After initially addressing letters to Patrick Geddes with "Dear Professor Geddes" during the first years of correspondence, Mumford switched to addressing his mentor with "Dear Master" after their first meeting in New York (Novak 183).

108 A generalist, as opposed to becoming "just another specialist" (Mumford, *Sketches from Life* 148).

109 Namely, that a "complete mode of living" is a worthy goal of existence; a conception of life "organically grounded and pursued with a little courage and audacity" that is embedded in a community as opposed to chasing after comforts and things as ends themselves (Mumford, *Findings and Keepings* 322–323).

110 In a piece entitled "The Disciple's Rebellion," Mumford confirms the largest influence Geddes had on him, both intellectually and practically: ". . . from the beginning it was in biology that he had made his most lasting contribution to my education – and even more profoundly to my life" (Novak 349).

Yet what impressed Mumford the most about his mentor was not Geddes's ideas per se, but his "audacious insurgency" (*Sketches from Life* 158). This injection of courage helped Mumford to "deepen and reinforce the foundations that other minds had already laid."[111] He was thus able to carve out his own historic structure and style. Drawing heavily on organic metaphors, Geddes's outlook was then later translated by Mumford into a full-blown historical analysis. The crossover from biology to sociology inspired by his mentor helped to emphasize the role of cities as containers and environments where humans must survive and flourish, and to advance the view of technology as an integral part of a vital "human ecology."[112] With the germ of that idea now in his mind, Mumford sought to write about the ecology and insurgency of human culture to map a future direction in the field of sociology.[113]

The second early and major influence on Mumford's education was economist Thorstein Veblen.[114] Drawing on Veblen's conceptual critiques of capitalism, Mumford could go beyond strictly objective representations of industrial development and build in more normative aspects into his descriptions of technology. In particular, Veblen's *The Theory of the Leisure Class* (1899) was vitally important to change the way his readers could view the relationship between production and consumption and for that matter, economies as a whole. Moreover, a Veblenian focus on institutions and their financial interests made it possible to show how business practices could trump technological progress in a dialectical relationship. From Veblen's *The Theory of Business Enterprise* (1904) Mumford had learned that there was sometimes an "underlying conflict between business interests and industrial interests" that overtly slowed technological advance (Mumford, *Technics and Civilization* 266). As another example of this influence, using Veblen's phrase the "ritual of conspicuous waste," Mumford explained how the dearth of military life and the power practices of ruling elites since the sixteenth century had enabled the emer-

111 One of those other minds, Charles Cooly, had written a monograph called *Social Organization* (1909) in which he employed an "organic view of history" that Mumford would later adopt into his own understanding of technological change (Mumford, *Sketches from Life* 158).

112 This notion can be found in editor Frank Novak's introduction to the correspondence between Mumford and Geddes: "Geddes' notion of a "human ecology" was important in shaping both Mumford's method of historical analysis and the scope of his interests" (Novak 25).

113 Suggesting that he and other writers "occupy the field" [of sociology], Mumford seeks Geddes's approval and thoughts on how to justify the importance of writing a history of civilization in the United States in a letter from 5 January 1924. Mumford intends to provide a diagnosis by historical analysis of the "things that have entered Western civilization during the last hundred years, and that threaten eastern civilization during the next hundred . . ." (Novak 189).

114 Mumford reveals the influence Veblen had on him both personally and intellectually in the "Acknowledgments" section of *Technics and Civilization*.

gence of modern forms of the machine, not only actively through war, but also indirectly by not waging it (96). In these times of respite, the focus has switched to the acquisition of luxury goods and the "consumptive ritual," leading to the construction of "non-utilitarian instruments" that have helped foster and advance important inventions (101). Awareness of institutional explanations of technology involving both direct and indirect factors contributing to technological change allowed Mumford to interpret the growth of industry and its entanglement with business interests and economic practices as a cultural phenomenon. And this is what made his first book on the history of technology rather unique.

Veblen's ideas about the relationship between technology and society laid out in successive books must have deeply affected Mumford's views of his own country and the need for economic reform after World War I. The fact that Veblen's analysis was set in America must have made it particularly credible to the young New Yorker. Theory turned into practice and to Mumford's good fortune, Veblen came to New York. The young student could then sit in on one of Veblen's courses at the New School for Social Research. By 1919, Mumford was able to have a more personal relationship with "one of his intellectual heroes" (Miller, *Lewis Mumford: A Life* 109). After having written a few articles for the publication, Mumford was asked to join the editorial staff at *The Dial*, where Veblen was already an editor. Thus, by reading his books and through personal relations at the same workplace, Mumford could enjoy the privilege of learning from what he would later call one of the "foremost sociological economist[s]" beside Marx and Sombart.

Their time together at *The Dial*, however, was to come to an early end. In 1919, the magazine focused on politics and society, but that was soon to change. *The Dial* had a socialist bent and as part of the reaction against what is known as the First Red Scare, the U.S. government unleashed a wave of suppression aimed at "radical" organizations. The magazine editors were forced to speak in front of a Congressional Committee and "cross-examined in secret for the alleged revolutionary activities of their magazine" (Miller, *Lewis Mumford: A Life* 111). Ultimately the magazine changed hands and Mumford and Veblen, along with a few others, were let go. Although Mumford would characterize the work of his "American elder" as somewhat "fragmentary and limited" in a 1925 letter to Patrick Geddes (Novak, *The Correspondence* 235), he later credits Veblen's entire oeuvre for making a "unique contribution to the theory of modern technics" (Mumford, *Technics and Civilization* 472).

The Convenient Fiction of Inevitability and the Embrace of Contingency

In a 1921 essay entitled "The Collapse of Tomorrow," Mumford described a feeling that emerged from the devastation to European populations and landscapes wrought by the First World War (*Findings and Keepings* 61–62). He might have also been describing his own career outlook. After being let go at *The Dial* and keeping he and his family afloat with various writing jobs, the young author-historian wondered what future path his writing career might take. A decisive shift occurred when he received advice from several friends that same year. As an author, he was told to play to his strengths of literary imagination and knowledge of sociology. In general terms this meant describing the subjects of his future writings both as they are and as they ought to be (Miller, *Lewis Mumford: A Life* 155–156).

Mumford wrote five major works between 1922 and 1931, none of them explicitly about technology. But no matter what the proximate subject of his writings, Mumford expanded the above framework to survey the past, appraise the present, and project a better future concerning one ultimate issue: the rise of the machine and mechanistic worldview. This reckoning with the Machine Age was taken up implicitly in his first few book projects and stayed with him for his entire career, prompting literary critic Van Wyck Brooks to remark that "Lewis was one of the few men, . . . who have not *ideas* but *an idea*, and he was to spend his life working this out" (Miller, *Lewis Mumford A Life* 163). Also underlying his work at this time was a sort of activist mode that he had learned while writing editorials at *The Dial*. Its guiding principle was "reconstruction," which meant that institutions could and should be "searchingly re-examined and reshaped on a more humane pattern" (Mumford, *Sketches from Life* 217–19). Instead of his colleagues' concentration on active efforts to pursue institutional and other changes in their various fields, Mumford, as a writer, would focus on a fresh understanding of America's cultural past and propose future values consistent with its base.

In what follows, three examples are taken from Mumford's first three books to illuminate an early version of his core historiographical methodology. Selection of this small sample size is intended to explicate a general historical and philosophical stance that is rather consistent from the beginning to the end of his career. This selection will show that even Mumford's early writings embrace a framework of contingency with respect to causality in history. He does not abandon this basic framework for his later works that use technology as a subject of historical inquiry. Another emphasis from these early writings was to retrieve a usable American past. Mumford wanted to use his literary and philosophical imagination to create a space for a type of history that would help Americans develop more organic culture (as opposed to mechanistic), and a connection to the past that would allow them to deal with the then post-war situation and positively prepare for the future.

According to Mumford, who was called to active duty during World War I as a Navy radio operator, all that seemed to be left from the "wreck of the Great War," were "wishes and dreams" (*Sketches from Life* 303). Inspiration for a reconstruction of dreams, he thought, might come from the historic presentation and examination of "underlying ambitions and wishes." Thus, he began his writing career not on a book about technology, but on one telling *The Story of Utopias* (1922).[115] Although a survey of utopias did not provide much guidance toward reshaping social and urban values in post-war America, it did allow Mumford to explore concepts that would emerge in his later writings about technology. For example, in a preview of his later, more inclusive understanding of technics and a rejection of the mechanistic mode of explanation, Mumford cautions against attempts to conceive of a good society merely through positive changes to its institutions. While discussing the weaknesses of nineteenth-century utopias, he argues that "externalism," or the idea that institutions were "external to man," did not sufficiently account for the people and environment that had created them (*The Story of Utopias* 247):

> It would be so easy, this business of making over the world, if it were only a matter of creating machinery. . . . These [nineteenth-century] utopias were so concerned to alter the shell of the community's institutions that they neglected to pay attention to the habits of the creature itself – or its habitat. That is why mechanical devices play such an important part, perhaps, in all these utopias (251–252).

Another example can be seen in his second book, *Sticks and Stones: A Study of American Architecture and Civilization* (1924), which attempts to "criticize the forces that from one age to another have conditioned . . . architecture, and altered its forms" (Acknowledgements). Mumford rejects overly deterministic explanations. Historical causes, he argues in *Sticks and Stones*, can be quite unpredictable. In a chapter assessing the co-development of architecture and civilization, the young scholar previews a theoretical stance he will later use to explain historical change in technology, one which involves a more contingent understanding of historical change.

> Looking back upon the finished drama, it is convenient to regard our community and our builders as creatures of their environment: once their choices are made, they seem inevitable. On this account even the pomp of the imperial architects can be justified, as the very

115 In the "Acknowledgements" section of *The Story of Utopias* Mumford credits Van Wyck Brooks for having suggested the project to him.

> voice and gesture of the period they consummated. Looking forward, however, this convenient fiction of inevitability is no longer serviceable: we are in the realm of contingency and choice; and at any moment a new factor may be introduced which will alter profoundly the economic and social life of the community (195).

His next attempt to understand contemporary American experience and culture in light of major mechanical and industrial change was his third book, *The Golden Day* (1926). In it, Mumford specifically treats "imaginative literature and philosophy as the key to [American] culture"[116] and duly makes use of the organic metaphor to critique the shortcomings of both looking to a European past and the philosophy of pragmatism[117] to confront the present:

> When we are integrated, we grow like the tree: the solid trunk of the past, and the cambium layer where life and growth take place, are unified and necessary to each other. If the pragmatists had read this lesson from history, they would not have sunk entirely into the contemporary scene; and if the pillagers of the past had realized this truth, their efforts to establish a background would not have been so superficial (230).

Read closely, the above passage suggests an early philosophy of history. Mumford later makes it more explicit by praising works that serve as fundamental stepping stones to frame a new philosophy which is "oriented completely towards Life as the dominant thought since Descartes has been directed towards the Machine" (282). Convinced of a biological interpretation of history, by 1926 Mumford had also read and reviewed Oswald Spengler's first volume of *The Decline of the West*, which he would later refer to as an example of recognizing the "biologizing of thought" that was occurring in modern thinking about history (Mumford qtd. in Cowley 205). In the review, he happily notes that Spengler's monograph "gives a final blow to the two misconceptions of history that guided the nineteenth century," namely its supposed "continuous linear development" and temporal and spatial eurocentrism (Mumford, "The Decline of the West" 368). Subsequent reviews of Spengler's work in 1929 and 1939 were highly critical, but in a letter to Geddes in late 1928, Mumford notes the "profound similarities – and similar profundities" between the Scottish biologist's and the German historian's thoughts (Mumford qtd. in Novak 278). One can only speculate what he meant by that statement, but Mumford may have been referring to the originality of their thought,

---

116 In the opening note to his 1926 book, Mumford defines civilization as a "material fact" and culture as its "spiritual form."

117 Pragmatism started as a nineteenth-century school of philosophical thought, mainly in the U.S., whose main tenants were to consider the practical consequences and uses of any claim. Its leading figures were Charles Peirce, William James, and John Dewey.

albeit taken to different ends, and the similarity of insights about the development of cultures whilst using biological terms.

Building upon the biological framework of culture set out before him, the pivot that Mumford himself ultimately performs in his first approach to writing explicitly about technology's role in history was to infuse his writings with the goal of a possible renewal of culture. More expansive than Geddes and in defiance of Spengler's ominous predictions about Western Civilization, Mumford was about to draft an alternative narrative to the blind technophilia or the prevailing social pessimism of the day. Instead of a desire to have his readers "moved . . . to devote themselves to technics instead of lyrics" like Spengler, Mumford asks if there was not a way to move on from that destiny, to choose to integrate the mechanical and the organic, while operating within a framework of contingency, to imagine a new world (Spengler 41).

Averting the Impotence of the Machine Age

The stock market crash of 1929 and the Great Depression that followed had caused a serious rethinking of societal organization and industrial values on the part of a few scholars. Exactly one hundred years after Etzler's *Paradise* appeared,[118] another book, *The Great Technology* (1933), attempted to apply a similar mechanistic concept of history to espouse "the idea that we have the capacity for setting up an economic paradise on earth" (Rugg 67). The author, educator Harold Rugg, was a contemporary of Lewis Mumford. Rugg viewed the crisis of the time as an opportunity to reconstruct American society along scientific designs in order to reconcile technology with democracy by applying the scientific method to "all of man's social problems" (18–19). The book's main thesis is that this "particular depression *marks the closing years of an epoch*," a time that is characterized by an "*undesigned and uncontrolled attack upon virgin continents*," and that resulted in "*expansion without design and control*" (15–16). Influenced by Veblen and by the Technocratic movement, which saw engineers playing a vital role in reshaping the economy, Rugg's prescriptive volume envisioned society at the crossroads of a coming age, which he called "the Great Technology." This new era, he was convinced, could be brought on by properly applying the increase in productivity of the Second Industrial Revolution by designing and implementing

---

118 Etzler's *The Paradise Within Reach of All Men* was published in 1833; Rugg's *The Great Technology: Social Chaos and the Public Mind* was published in 1933.

a program of reform to be carried out by "scientifically oriented men from many professions" (105).

Rugg's book and ideas can be criticized for being overly optimistic and vague, but it is his usage of the word technology to denote a new epoch that is of present interest. In doing so, Rugg conflated history, historiography, and technology to revive a future-oriented, similarly utopian and romanticized twentieth-century version of Etzler's initial dream. He thus returned to the technical utopian idea that technology, understood here as technical decision-making based on the scientific method, could drive history along a lawful, linear path towards the goal of a more just society. Despite being a slightly more modern and modified version of technical utopianism, Rugg's program exhibits all the hallmarks of a framework of necessity in history. Certainly, the projected inevitability of social change brought on by noble-minded technicians can be criticized in his work. Yet in a small way, his book may have got one thing right. In a section entitled "the impending collapse is already stirring thoughtful people," Rugg unwittingly foreshadowed a book, Mumford's *Technics and Civilization*, which would not only be central to a contingent turn in the historiography of technology, but would also later be considered a major classic in the field (Rugg 25). This may be in part due to Mumford's ability to separate elements of utopia, as he had done in *The Story of Utopias* (1922), from technology, as he would do in *Technics and Civilization* in 1934. What also made Mumford's account of technology in history different is that he stretched his narrative to describe a much larger span of time. This longer view, along with his depth of scholarship, are the main reasons why his account has survived, whereas many other mainly prescriptive accounts of progressive and positive technological change, including the technical utopias written by Etzler and Rugg have become obscure.

In an unpublished manuscript of a first book draft in 1930, Mumford had mused about a solution to the "present dilemmas," concluding that the situation called for a "new philosophical idea" because the problems of modern society could not be limited to "specific attacks on specific evils and specific points of danger" (*Findings and Keepings* 200–09). He wanted to rethink the basis of modern life and thought, foremost its ideals, embodied by Soviet Russia and the United States, which he claims adhere to the "mechanistic plan of life" and "absolutist ideology of the Power State." A new orientation towards life would have to come in many forms and would not be quickly accomplished. It was, as he said, "not a task for the next ten years, but for the next two hundred." To help him accomplish this task, he began his research not by looking ahead to the next two hundred years, but by taking on the role of historian once again, and by looking backwards.

Mumford took a comprehensive approach in his first major attempt to understand the relation between technology and society. Acknowledging that there had been other attempts to write a history of the machine before his, Mumford notes in the bibliography of his 1934 book that, in his opinion, there had not been a "single, comprehensive and adequate history of technics in English" (*Technics and Civilization* 448). He had read Harold Rugg's book and praised its value for suggestions regarding a concern with the "educational problem of realizing the values of modern industry and of controlling the machine" (447–448). However, Mumford laments both Rugg's book and Stuart Chase's *Men and Machines* for their "limitation of historical foreshortening" (468). More of an author-historian than an educator like Rugg, who proposed a nation-wide campaign in hopes of "arousing the public mind to the great alternatives" (Rugg 20), Mumford thought that a change of mind could not be achieved without a long historical overview of the rise of the machine in Western Civilization. In his opinion, up until the publication of *Technics and Civilization*, the only adequate history of technics was *A History of Mechanical Inventions* by Abbott Usher. But he comments that, while Usher's account is critical and exhaustive in certain areas, it does not cover "every aspect of technics" (*Technics and Civilization* 448). It was indeed the broad scope and interpretive value of his inquiry into the history of technics that made Mumford's book unique.

In an article published in 1930, Mumford integrates the various strands of thoughts on the matter into a coherent political philosophy. Articulating the development and theoretical basis of his "elementary political faith," Mumford argues for egalitarian distribution of basic needs or "basic economic communism" that values an orientation towards life and leisure time over one which overly embraces things and work (*Findings and Keepings* 318–320). This insight, he hopes, will help society avoid what he sees as the impotence of the Machine Age, meaning that "comfort and not life" is the ultimate goal. This includes the notion that "elementary conditions of existence" are an "end" and not a right (*Technics and Civilization* 400). Such a recognition that each human being requires nearly equal amounts of basics for life like air, water, clothing, food, and shelter, in his opinion, should result in social institutions being adapted to establish a minimum basis of life. It would allow people to pursue the "chief business of life" – "growth, reproduction and development" – as opposed to embracing an attitude that seeks to subordinate fostering life-affirming interests to the production of new technologies that only provide comfort and ease (400). Historical framework and political philosophy in hand, Mumford set out to write what amounted to be the first of four books in the so-called "Renewal of Life" series. The project started as an invitation to teach a course at Columbia University on the "Machine Age in America" and was enriched by a four-month research trip to the technology museums of

Europe funded by a Guggenheim fellowship (*Sketches from Life* 448–449). These experiences ultimately helped Mumford broaden the scope of inquiry he was taking to consider a "larger idea [that he had thus far] avoided tackling."

Technology as a Historical Subject – Personality Affects Form

One of the goals that Mumford stated in the final edition of his next book was to instruct his readers as to the origins of the "radical transformation of the environment and routine of life" (*Technics and Civilization* 3). This altered mode of life or pattern of living is the larger idea that had eventually become the object of the study. Initially, Mumford was more interested in this end result and how to create an alternative future: he wanted to describe the mechanical mode of existence in modern times and how it might be changed to assimilate the machine toward an "organic ideology."[119] Yet what he realized during the writing process is that the object of the study needed a subject: historical particularities and agents that effected those changes. In order to gain perspective on the origins, he began exploring one specific area of his research, an area that eventually became the subject of his writing, which had outgrown and outpaced the term industrial arts and the mere application of science to include technical heritage. By tracing it holistically in history, Mumford thought that he could better understand how past expression of the human personality affected the current form of civilization.

Technics and Civilization represents Mumford's first attempt to embed the history of technology into a larger cultural framework of human civilization. In an introductory note, the author explained that the initial draft of the book was much larger in scale but had to be limited due to what he deemed as the necessity to expand on the technics section. It seems that in writing this small caveat, Mumford wanted to prepare his readers for the radical and unconventional reinterpretation of the history of technology that followed. Introducing the notions of "long internal development of technics" and "change of mind," Mumford began his argument by laying out two ultimate causes in the development of the so-called Machine Age: cultural preparation and a reorientation of desires, habits, and purpose (along with the external, conventional and proximate causes: the institutions, machines, and technics themselves) (*Technics and Civilization* 3–4). To make sense of the development of the machine, it was important, he argued, to "explore in detail the preliminary period of ideological and social preparation" in

119 These and other topics are largely addressed in the last two chapters of the book: "Assimilation of the Machine" and "Orientation."

order to understand the "dominating role played by technics in modern civilization." The latter, he argued, is what differentiated "us" from "them." It is an alterity that he deemed particular to Western Europe[120] – the adaption of the mode of life to the machine. Not limiting his history to the "existence of new mechanical instruments," the author intends to dissect the culture that was "ready to use them and profit by them so extensively."

Contesting the narrative of societal progress by means of technology, Mumford viewed the benefits of the machine as somewhat stunted in modern times and he lamented the "loss of form" and "impoverishment of life" that has resulted beside increasing technical advancement (5–6). He did not view this as the inevitable outcome of a lawful sequence of events, but rather that of contingent factors that could not all be objectively reproduced. He insisted that technics do not form an external and independent system, but were part of culture, and therefore subject to human and environmental factors. One of the main objectives Mumford stated at the outset of the book was to emphasize "human choices, and aptitudes and strivings, deliberate as well as unconscious, often irrational when apparently they are most objective and scientific" as causal, yet unpredictable factors in historical outcomes.[121] Mumford hoped that his readers could be part of a reorientation and transformation of values. Ultimately, he believed that the machine could be integrated into society and its use mastered for human purposes.

Technics and Civilization – A Close Reading

The form of historical explanation (discursive argument) employed in *Technics and Civilization* has been described as organicist[122] and moralist.[123] In a paper locating

120 In a critique of his own culture, Mumford argues that other cultures, like the Chinese, the Arabic, the Greek "had machines; but they did not develop 'the machine' " (*Technics and Civilization* 4).

121 In the last section of the last chapter of the book, Mumford restates his commitment to the unpredictability of history: "The issue is not decided: the results are not certain: and where in the present chapter I have used the prophetic form I have not been blind to the fact that while all the tendencies and movements I have pointed to are real, they are still far from being supreme: so when I have said 'it will' I have meant 'we must' " (*Technics and Civilization* 434).

122 As Hayden White explained it, the narrative structuring employed by the organicist mode, "attempts to depict the particulars discerned in the historical field as components of a synthetic process" (*Metahistory* 15–16).

123 In her contextualization of Lewis Mumford as a historian of technology in *Technics and Civilization*, Rosalind Williams states the "well-known fact" that Mumford was a moralist. As I interpret her use of the word, the term can oscillate between neutral and slightly negative connotations, or

the central theme in Mumford's writings, Leo Marx viewed the "conflict between organicism and mechanism" as the "chief ordering principle" that permeates and guides his writings ("Lewis Mumford: Prophet of Organicism" 172–174). It is an argument that led Marx to describe Mumford's historical authorship as infusing a "disembodied abstraction," or the idea of mechanism or the machine, with historical agency. Marx is surely right in classifying what he called an "analytic mode" as being "ahistorical." This is because the description at least partially fits Mumford's early approach to the history of technology since he tended to see the individual entities that make up technics as components of a historical process that constitute the machine, or the mechanical mode of life. In other words, an abstract entity like the machine is used to synthesize the desperate components of technics into an integrative whole. Scholar Rosalind Williams's characterization of Mumford's organizing concepts in *Technics and Civilization* as moralist is also accurate. As she explained in detail in a chapter on his methodology as a historian of technology, Mumford needed to use moralism to structure the vast amounts of information that he had gathered. When reading through the book, it becomes rather obvious both in the more descriptive first chapters and the more prescriptive later chapters that he had clear moral preferences as to what constituted a correct and an incorrect application of technics for a more balanced life within machine civilization.

The current analysis attempts to go beyond these characterizations and emphasize Mumford's analysis of technology as contextualist[124] and radical.[125] A first step in this direction is recognizing that constructing an argument of historiographical explanation in the organicist mode is a narrative choice. Mumford, as an author-historian, could employ this mode, as well as others, to explain and ultimately interpret the past on different levels. Having one mode like organicist operating on a macro level does not preclude having another mode like contextualism operating on a micro level. Following this view, part of the authoring or emplotting process using the organicist mode to represent the past necessarily involves the imposition of a model or concept onto atomized facts and evidence. Yet

meanings between someone being an arbiter of proper or improper behavior and someone who rather audaciously imposes their ideas about conduct onto others ("Lewis Mumford as a Historian of Technology in *Technics and Civilization*" 44).

124 Hayden White defines contextualism as being informed by the idea that historical entities "can be explained by being set within the "context" of their occurrence," to link together threads of history "in a chain of provisional and restricted characterizations" to show significance (*Metahistory* 17–21).

125 According to Hayden White, a radical ideology is generally characterized as believing in the "necessity of structural transformations" to reconstitute "society on new bases" (*Metahistory* 24–28). This ideology fuels the recommendations made in the last chapter of *Technics and Civilization*.

when it comes to the actual truth concepts that underlie a particular understanding of technics at a specific time, Mumford also made use of the particular historical context in which certain trends appear to map them onto a larger perspective of the past. A second step in the argument concerns the characterization of the structuring of the book in moral terms instead as mearing being radical. Accused of writing about ideas rather than facts, Mumford's attempt to gain a better understanding of technology's past and present in order to be able to reorient future society according to different principles is part of this tradition. It is a tradition that stands in opposition to the more accepted conservative or liberal ideologies in historical studies. Mumford used the concept of the machine as a subject to project continuity onto historical time that can be characterized as the *present past*. As a consequence, his account of history roots the present in the past and allowed him to play the role of moral prophet on mechanized civilization by advocating for a more organic and imminent future, one which is contingent on human choices. In what follows, Mumford's ideas about technology and technological change are shown to exhibit these two characteristics through a close reading of the book.

Mumford opens *Technics and Civilization* with a chapter entitled "Cultural Preparation" that serves as the basis for the sequential phases of history examined in chapters three through five. As Mumford pointed out in a 1933 letter to his trusted friend, Henry A. Murray, he had to "deepen and expand" his original chapter to an entire book and include the seven centuries before the so-called Industrial Revolution, which he thought served as a preparatory period that made it and any subsequent developments possible.[126] In a later chapter, he emphasizes the salience of this *longue durée* of cultural preparation to a more sophisticated understanding of the machine in history:

> So far from being unprepared for in human history, the modern machine age cannot be understood except in terms of a very long and diverse preparation. The notion that a handful of British inventors suddenly made the wheels hum in the eighteenth century is too crude even to dish up as a fairy tale to children (109).

In describing preparation for the machine, Mumford displayed a willingness to situate causality on multiple levels and in less linear relationship to specific or predictable outcomes. In the first section, Mumford delineates the differences be-

126 In a letter to Henry A. Murray dated 12.02.1933, Mumford wrote: "The point is that the whole change of mind, which made possible the developments of the last two hundred years, was already pretty well accomplished by the end of the seventeenth century" (*"In Old Friendship" The Correspondence of Lewis Mumford and Henry A. Murray, 1928–1981* 82–83).

tween tools, machines, utensils, apparatus and utilities,[127] but more importantly for his later arguments, between these objects and *the machine*, or the "entire technological complex" (11–12). While he does expand the understanding of technology to include containers in the introduction of the book, as he admits himself, he did not trace this distinction through the rest of the book.[128] Instead he starts by describing the contingent "convergence of numerous habits, ideas, and modes of living" foremost being the "routine of the monastery," that proximately led to the mechanical conception of time and then ultimately to the "abstractions of measured time and measured space" (12, 22). In doing so, Mumford demonstrated an awareness of the latent determinism implicit in such a mode of thought, that being the reliance on abstractions and numbers to understand the world and to project future outcomes.[129] As a writer that mixes different modes of historical explanation, he saw a danger in deriving meaning completely from time and place.

The next section focuses on the individuals who control the development and "style" of the machine within an economic system, and according to their social roles. The influence of capitalism and the monetary system, he argued, "conditioned" and "reacted upon" technics and vice versa, in the process, changing the "economy of needs" to an "economy of acquisition" (24–28). This "quest of power by means of abstractions" included increasing demands on workers by the capital-owning feudal families and merchant classes for their own private benefit. Technologies were not used for the promotion of the general welfare of society, but rather, as Mumford puts it, "for the aggrandizement of the ruling classes." Such fundamental criticism of systemic realities mirrors Veblen's work because Mumford assigned blame to the actors contextualized within larger pecuniary and power systems. Veblen's influence also figures prominently in Mumford's critique of capitalism in the last chapter of the book.

Other changes that helped form a "complex social and ideological network" that set the stage for the Machine Age were claimed to be the move from stories

127 A tool has flexibility of use and requires a skillful operator and a machine performs a specialized function. Interestingly enough, Mumford refers to a machine as a "sort of minor organism," thus at least metaphorically conflating mechanism with organism in his opening pages. Utensils can be simple holders for objects like boxes or pots, while apparatuses are more complex containers that include vats and kilns. Finally, utilities, like the roads and aqueducts and buildings, are specialized conductors to be used for specific functions (*Technics and Civilization* 10–11).

128 Mumford laments this lack of follow through in a review of his own book 25 years later ("An Appraisal of Lewis Mumford's 'Technics and Civilization' (1934)" 529).

129 "In time-keeping, in trading, in fighting men counted numbers; and finally, as the habit grew, only numbers counted" (*Technics and Civilization* 22).

to facts and an interest in nature at the expense of animism, magic and the human body being part of a mythic whole, the increase in social regimentation through Protestantism, and the ethos and practical application of the scientific method.[130] Amongst these multiple cultural causes, two resulting conflations were clear to Mumford, technics with religion and work with life. The first involved the replacement of the immaterial, infinite, and eternal god with material, finite, and planned obsolescent ones: "mechanics became the new religion, and it gave to the world a new messiah: the machine" (45). The second meant a reduction of life to work: "To live was to work: *what other life indeed do machines know?*" (53).

In what can be described as a methodological homage to his mentor, Mumford devoted the entire second chapter to delving further into the search for specific ideal type "agents of mechanization" through use of a scheme that profiles a conceptual valley from top to bottom, an idea he borrowed from Geddes (74). Instead of focusing on how economic factors influenced technics, he attempted to demonstrate how various occupations, like the miner, the hunter, and the woodsman can be matched to locations in a mountain-and-river system and foster a sort of occupational mentality that contributes to the subsequent developments of the machine. These include, for example, the miner's practice of exploiting the environment to "make his pile," the similarities in ideology between a hunter's quest to eliminate rivals and secure game and the soldier's profession of hunting other men, the abstract "absentee ownership" and monetary goals of the financier, and the "productive drive" of technics and science and the "consumptive pull" of the courtly lifestyle (74, 77, 102–106).

Mumford, as an author-historian, characterized the historical development of the machine with a clear direction. This framing of his account is accompanied by sources of change, some of which are particular, some of which are generalized. Taken together, Mumford described such forces and idealized personalities, their tools and their work environments as affecting the direction of historical reality. On a more ideological level, the cultural preparation for the machine and the agents that brought about mechanization contributed to a cultural character, which in turn shaped larger social patterns and meanings.[131] According to this view, civilization is an organism, and certain dominant aspects of its personality affect its historical form. Mumford's first theory of history is thus conceived as a

---

130 Mumford neatly summarizes these cultural causes of mechanization as follows: "clock-making: time-keeping: space-exploration: monastic regularity: bourgeois order: technical devices: protestant inhibitions: magical explorations: [. . .] the magistral order, accuracy, and clarity of the physical sciences" (*Technics and Civilization* 59).

131 "The machine cannot be divorced from its larger social pattern; for it is this pattern that gives it meaning and purpose" (*Technics and Civilization* 110).

human-driven world in which order and meaning are provided first through cultural practices and rituals, which include technics, but also through interaction with institutions and material realities. The outcome of the entire process, when one understands the ideas in *Technics and Civilization* as final, is both predictable and inevitable – the loss of form due to a malignant personality.[132]

Commenting on Mumford's "singular innovation and enduring contribution" to the history of technology, Rosalind Williams indicated Mumford's "insistence on technics as an expression of human personality" ("Classics Revisited: Lewis Mumford's *Technics and Civilization*" 140). Instead of a list of inventors and inventions that have "changed or improved society," Mumford's final draft of the book provided a new and different interpretation of the development of the machine and machine civilization, combining internal mind and external technics to explain technological change. But Williams noted that, "the book we know as *Technics and Civilization* was an accident. Mumford intended to write a book, not about the technological past, but about the cultural future. He planned a brief discussion of machinery as a preface to a far more comprehensive cultural study, to which he had given the working title 'Form and Personality' " (141).

This anecdote about the actual writing process reveals how Mumford's ideas changed during his research stay in Germany. Although he retained an organic mode of explanation into the end phase of the project to maintain a specific narrative – his overarching ideas regarding where technics, the machine, and the current mechanical civilization had been, where they are now, and how they would steer future society – he did not just synthesize and abstract ideas as an organicist author of a narrative fiction. He contextualized historical facts that he had dug up, most notably in the library of the Deutsches Museum in Munich.[133] It seems apparent that Mumford profited from the detail of German historical accounts and access to the German discourse of *Technik*, a cognate of technics, and its close connection to discourses on *Kultur* (culture). In the bibliography of *Technics and Civilization*, Mumford singled out Werner Sombart's *Der Moderne Kapitalismus* as a "monument of twentieth century scholarship" (448). As Eric Schatzberg pointed out, Ger-

132 In a letter to his friend, Mumford claims the opposite: "[Some] have hastily jumped at conclusions that are not warranted by the book, because they took the statements in Technics as final, instead of preliminary, to be qualified, in turn, by what I write about the group and the personality" (*"In Old Friendship" The Correspondence of Lewis Mumford and Henry A. Murray, 1928–1981* 91).

133 Mumford underscores the importance of his historical research at the library of the Deutsches Museum in which he had access to "a growing literature on the historical development and cultural transformations of technics" ("An Appraisal of Lewis Mumford's 'Technics and Civilization' (1934)" 528).

man influence on the book may have come in the form of Mumford's "humanistic" use of the term technics, in which the concepts of technics and culture are part of an overarching *Geist*, or "human mind and spirit" (*Technology* 150). When combined with geographic and economic archetypes he had learned from Geddes and Veblen, *Technics and Civilization* became a polyglot reach into the past and a composition of German, British and American ideas.

Part of his mapping out the mechanical past includes a periodization of eras, or as he called them, "successive waves," which he thought had gathered "strength and momentum[134] as other institutions in civilization were weakening" (*Technics and Civilization* 4). This choice of metaphor is instructive as it represents a break with a strictly linear, progressive understanding of time in history, synthesizing Greek (cyclical) and Christian (linear) concepts of temporality, to produce a hybrid form. Moreover, his suggestion that the third of these so-called waves was moving in a different or even opposing direction to the other two receding ones confirms at least a partial commitment to a contextualist[135] understanding of historical causality and technical development (5).

Utilizing this mode of explanation, the next three chapters of the book present a subdivision of the development of the machine and machine civilization into three phases: the Eotechnic, the Paleotechnic and the Neotechnic.[136] These three phases, as Williams rightly pointed out, also correspond to the "three-act drama of Life – equilibrium, breakdown and renewal," which is a prominent theme in his writings ("Classics Revisited: Lewis Mumford's *Technics and Civilization*" 143). Williams emphasized the "confusion" this overriding moralist scheme causes in Mumford's representation of historical change, which she called "organic determinism." Clearly, the historiographical practice of periodization can appear rigid. It opens an author-historian like Mumford to the critique of squeezing a large number of flowing and circular changes to fit between the arbitrary brackets of a chronological past; of being too compatible with a more deterministic understanding of historical change. In the particular case of *Technics and*

---

134 More recently, the concept of technological momentum has been championed by Thomas P. Hughes. For a clarification of the term, see "Technological Momentum" in *Does Technology Drive History? The Dilemma of Technological Determinism* (1994).

135 According to Hayden White, the contextualist mode of explanation attempts to *relatively integrate* "phenomena discerned in finite provinces of historical occurrence" in terms of epochs and the "flow" of time is described in terms of a wave-like motion (*Metahistory* 18–19).

136 The first phase, or Eotechnic, was Mumford's unique contribution to an understanding of the development of the machine that he described as the "important period of preparation" for the latter two phases, which he had borrowed from Geddes (*Technics and Civilization* 109).

Civilization, the trifurcated structure overshadows and contradicts the more contingent and cultural causality of the first two chapters.[137]

As for the consequences of technological change, the three "phase" chapters are bookended by two words that reveal the figurative arc of development that Mumford fears humans were taking technology. He begins the chapter on the Eotechnic phase using the term "technical syncretism" to describe the at times random "wind-blown seeds" or fragments of technics that are transported to new cultures (107–110). These parts are drawn from any number of sources and compiled into technics over a long preparatory period of cultivation in a search for equilibrium. Free, yet contingent, these were the origins of the modern machine age. In contrast, Mumford ends the chapter on the Neotechnic phase with an appropriation of a term he borrowed from Oswald Spengler, the "cultural pseudomorph" (263–267) This word demonstrates the "miscarriage of the machine" that Mumford articulates in the book. In the "present pseudomorph," the current form of civilization, dominated by the values of mechanization and power as ends in themselves, was not changing toward more organic forms, but merely being supplied by different materials, increasingly mechanized inventions, and new technics. In other words, new technics were being used to uphold old structures. There were political and financial interests in these structures resulting in technological inertia and Mumford viewed this as a breakdown in civilization.[138]

In *Technics and Civilization*, Mumford employed both traditionally historical (realist mode) and literary modes of explanation. This latter observation may have led Williams to summarize that Mumford "writes not so much as a historian of technology [but] as a retrospective heuristic utopian who connects changes in values with changes in technology" ("Lewis Mumford as a Historian of Technology in *Technics and Civilization*" 65). Yet this characterization primarily applies to the aforementioned overall narrative emplotment of the book and not to his detailed analysis and contextualization of the agents, practices, sources of power, and materials that all interact to effect technological change. As she acknowl-

137 In a candid and ironic third-person assessment of the limitations of breaking up time into technical phases, even if they were never intended to be definite, Mumford congratulates himself "over the fact that the division he used never effectively caught on; and it broke down in his own mind before it could do any serious damage" ("An Appraisal of Lewis Mumford's 'Technics and Civilization' (1934)" 532).

138 Mumford asks the question to what end technics should be applied: "For what has been the total result of all these great scientific discoveries and inventions, these more organic interests, these refinements and delicacies of technique? We have merely used our new machines and energies to further processes which were begun under the auspices of capitalist and military enterprise: we have not yet utilized them to conquer these forms of enterprise and subdue them to more vital and humane purposes." (*Technics and Civilization* 265).

edged, Mumford is both prescriptive and descriptive in his assessment of technological change in the book. When he tried to fit facts to his scheme, his writing exhibits elements of a framework of necessity in history. These are often seen in romanticized passages like this one on wind and water as sources of power:

> Thanks to the menial services of wind and water, a large intelligentsia could come into existence, and great works of art and scholarship and science and engineering could be created without recourse to slavery: a release of energy, a victory for the human spirit. Measuring the gains not in horsepower originally used but in work finally accomplished, the eotechnic period compares favorably both with the epochs that preceded it and with the phases of mechanical civilization that followed it (*Technics and Civilization* 118).

This framing can also be observed in generalizing summations that are supposed to be universally valid for a particular culture like the following description of Europe circa 1750: "The great dilation of the senses, this more acute response to external stimuli, was one of the prime fruits of the eotechnic culture" (150). In such passages, it is rather apparent that Mumford's mode of explanation is that of a storyteller, exhibiting few similarities to a historian in the realist mode. The "confusion" that Williams referred to results from these storytelling elements, like when he appears to attribute changes in society to the introduction of coal to supply mechanical power: "Out of this coal and iron complex, a new civilization developed" (156).

Still, the variety of contextual and contingent factors that play into technological change included in Mumford's first major attempt at a history of technology should be underscored. In a section on the primary mechanical inventions of the Eotechnic phase, Mumford reminded his readers that changes in technology are always part of a broader system of causal inputs, which include the environment in which they emerge. Alongside the printing press and mechanical clock, he wrote about the importance of paper as a material, "social inventions," institutions like universities, medical schools, the scientific academy, the industrial exhibition and the laboratory, and finally the containers, a category which includes the creation of the factory (137–139). He even went as far as to mention the importance of manners in technological change, writing that "energy is not merely a question of bare physical resources but of their harmonious social application."

Central to Mumford's attempt at understanding the development of technics was specifying the interrelationships within a so-called technological complex.[139]

---

139 Mumford explains that each phase forms a "technological complex," meaning that it: (1) "has its origin in certain definite regions," (2) "tends to employ certain special resources and raw materials," (3) "has its specific means of utilizing and generating energy," (4) possesses "special forms of production," (5) generates "particular types of workers" and their training, (6) maintains

Underneath a macro-level grouping of technological changes in a three-part scheme, he provided a particular and multi-causal analysis of change within the context of that specific phase. With the exception of the "mutation" of the so-called Neotechnic Phase, most of the arguments in the book point to a continuity in technological development, not a complete break with the past. Once a new practice, a new raw material or new way of generating energy has been introduced, Mumford viewed it as a change in regime, always varied and contingent on the context. In this case, the context was human ecology, which consists of many moving parts and is always subject to change. As an example, Mumford used this long-view of technological development to suggest an answer to the historiographical quirk (in English language sources) of assigning a linear interpretation to the Industrial Revolution as the single cause of the Machine Age – England's lack of participation in the technics that were part of the greater cultural preparation for the machine.[140]

While the macro-level groupings and storytelling elements remain problematic, the micro-level historical analysis exemplifies a move away from a framework of necessity in history when explaining technological change. Two passages stand out in particular for their direct characterization of his views on this matter. In the first, Mumford denounced the narrow-mindedness of believing in universal technological progress: "This picture of a steady, persistent, straight-line, and almost uniform improvement throughout history had all the parochialism of the eighteenth century" (182). The second gives a definitive answer to Mumford's position on the question of whether technology is the driving force in history:

> for the necessary social institutions and the explicit social purposes requisite even for complete technological fulfillment are lacking. The gains in technics are never registered automatically in society: they require equally adroit inventions and adaptations in politics; and the careless habit of attributing to mechanical improvements a direct role as instruments of culture and civilization puts a demand upon the machine to which it cannot respond. Lacking a cooperative social intelligence and good will, our most refined technics promises no more for society's improvement than an electric bulb would promise to a monkey in the midst of a jungle (215).

"certain aptitudes and discourages others," (7) "draws upon and further develops certain aspects of their social heritage" (109–110). He categorizes the three phases as follows: (1) the eotechnic phase as "a water-and-wood complex," (2) the paleotechnic phase as "a coal-and-iron complex," (3) and the neotehnic phase as "an electricity-and-alloy complex" (110) (*Technics and Civilization* 109–110).

140 Mumford explains this reasoning in a chapter entitled "England's Belated Leadership" (*Technics and Civilization* 151–153).

In his first major book on technics, Mumford had the acumen and moral courage[141] to write about more than only progressive, scientific, and successful technics and technological change. His sections on "the destruction of the environment," "the degradation of the worker" and "the doctrine of progress" show him at his very best, critiquing aspects of the Paleotechnic phase for their callous disregard for plant, animal, and human life. This focus helps expose the dark underbelly of so-called societal advance by means of technology while addressing the ignorance surrounding issues of technological destruction and inefficiency. It also highlights the grave costs of blindly adhering to the "cardinal doctrine" in modern times (182).

Throughout the book, the critical voice of a cultural historian of technics is accompanied by one belonging to a human ecologist of mechanical society. After having rendered a historical account of the "transfer of order from God to the Machine" through the first five chapters of the book, Mumford's sociological analysis becomes increasingly relevant to the modern reader in the final three chapters (34). He began the sixth chapter with such sociological analysis by describing some of the characteristics of the presently dominant religion: machine civilization. He viewed the machine itself as "ambivalent," meaning that the historical record was mixed and the end results of its changing influence in society are ongoing and inconclusive. It has both served and hindered liberation; it has both expanded possibilities and repressed rights (283). In particular, Mumford strikes a somewhat objective tone as he exercised both support and critique for things such as "the regularization of time, the increase in mechanical power, the multiplication of goods, the contraction of time and space, the standardization of performance and product, the transfer of skill to automata and the increase of collective interdependence" (281).

In contrast to this seemingly neutral wording, Mumford's moral assessment of the (then) current situation was indeed conclusive. He lamented what he described as a "perversion" of the machine in the "translation of technical improvements into social processes," the result of which tended toward absolute power and social control by the ruling classes that have organized, controlled, and owned the machine (281). As he continued, the lack of "social evaluation," meaning the "failure to evaluate the machine and to integrate it in society," was to ignore the need for a new set of values in tune with life processes (282–283). He was warning his readers against the possibility of what he would later call "authoritarian technics" becoming a creative force of its own that could order a new envi-

141 The German word *Zivilcourage*, literally "civil courage," is more apropos in this case as it emphasizes the non-violent means and personal risk taken in attempting to write opposing dominant social norms, like an exploitive economic system or one-sided view of technical progress.

ronment without democratic control. He sought a different way forward through the last two chapters of the book. Chapter seven is a historical account of the assimilation of the machine into society and chapter eight is a radical list of proposals to redirect/rebalance society more toward the organic.

Moving past his assessment of modern machine civilization, in these final chapters Mumford returns to the role of cultural historian. He waded through the various compensations, retreats, and reversions that individuals and groups have undertaken and undergone with regard to the machine. All the while his focus lies not only on the products of the machine, but on the place of man in a mechanical environment and the active processes that shape it. Neither of the two is seen as final or determined, and in this way, he was able to underscore the dynamic and contingent role human culture plays in shaping technics. Ultimately, he saw human society presented with an ongoing choice between the extremes of "co-operation" of machines and "slavery" to the machine (278). He rejected solutions to the current relations between technics and civilization that included direct attacks and reactions that are too utilitarian or romantically one-sided.[142] The latter of which, he observed, while being an "important corrective" to the machine, had had negative historical consequences when taken to an extreme. Regression into a retrospective and romantic view of history had resulted not only in a metaphysical break between the present and the "immediate past" in favor of a romanticized and fetishized past, but also in a similar disruption between the present and the future (288, 303). According to Mumford, a "neurotic retreat" into either the historical space of an absolute past (292), or the physical return of humanity into space called "nature," meaning primitive or rural conditions (299), conceded victory to the machine and represented a defeatist approach to the problem. Making an adequate response more difficult represented a similar retreat into primitive forms, which, when unchecked, not only intensified a polarity between organic and mechanical forces, but also hindered the transition from more "historic forms of capitalism and . . . the machine" to a more "life-centered economy" (302–303).[143] In this last statement, Mumford shows his radical critique

142 Utilitarian thought, generally speaking, would ignore negative effects brought on by the blind pursuit of technology and profit while Romantic reaction to the machine, often ignoring the positive effects, took many forms. The three dominant ones that Mumford mentions are the cult of history and nationalism, the cult of nature and the cult of the primitive (*Technics and Civilization* 285–287).

143 Mumford was well aware of the effects of media. An example of a retreat into the primitive and the inability to transition into a new economy is magnified by the misuse of media, like radio or internet. A passage in *Technics and Civilization* foresaw the negative capacity of those in power at both the time of the book's publication in the 1930s (Hitler) and any current of future so-called leader. "Mechanical instruments, potentially a vehicle of rational human purposes, are

not only of responses to the mechanical environment, but of some of society's sacred cows.

Two other cultural practices followed. Further compensations that have ultimately been unsuccessful in their bid to generally upend the influence of current technics on the economy and society were the practices of sport and war. In dissecting the difference between more organic and universal play and the regimentation of sport in mechanized society, Mumford criticized the vicarious participation in acts of bravery and bravado through mass sport at first as limited. Strongly geared toward spectacle, Mumford supposed that it provided the crowd only a superficial release from mechanized life through the "illusion of being completely mobilized and utilized" toward a collective cause, merely a win (305). He later admonished the technique of mass sport as not being true play, but a profit-making, means-to-an-end business that affected many other sectors. In his opinion, society was actually harmed by participation in spectator sports. Beyond temporary relief of otherwise suppressed emotions, their practice contributed to the strengthening of the machine and diverted energy away from more fundamental aspects of life. In the following section on the "cult of death," a distinction was made between sport, which is staged, and war, which is real, to emphasize the extreme position occupied by war as a form of conflict resolution, especially in mechanical society. The goal of war is no longer resolution of who is better, but the elimination or full submission by force of the opposing person or group (308). For Mumford, war represented the "supreme drama" of mechanized society (309). It was a tragic theater that served both as the source of mechanism, but also as a primitive compensation for people who could not respond to conflict with an understanding of others' values and culture.

Mumford continued his analysis of compensations to the machine by criticizing what most people might consider their hobbies. He wrote about an escape into the "cult of antiquarianism," by which he meant anything from the past, or its opposite, embracing change solely for change's sake, or even using films or literature as a "mere vehicle of withdrawal" (315). He lamented that these behaviors had all ceased to be spontaneous, private, and free, and had become part of the "amusement business" and thus a vested interest of investment. Once again, Mumford demonstrates an emphasis on humans and their ecology when showing the "anesthetics" and "shock-absorbers," as he called them, that people use to combat the pervasive psychological and physical stress in a mechanical environment (316). He added that they not only help one compensate, but also prepare

scarcely a blessing when they enable gossip of the village idiot and the deeds of a thug to be broadcast to a million people each day" (303).

one for more "disastrous forms of release," like the real shock of war. Part of his radical interpretation of history is that he has his mind on structural changes in institutions and values, not on the complete replacement of these forms of outlet. But what response could be effective and what would be the driving force to enact effective change, as he asked, if the products of the machine did not "further life and enhance its values" (318)?

Instead of succumbing to the negative potentialities of the machine, Mumford reminded his readers of the multi-directional nature of change between humans and their environments. Invoking both necessary and contingent forces in historical change, Mumford drew on his mentor's biological interpretations to emphasize humans' ability to harness their own energy and capabilities to effect change in a greater human ecology through the arts, which included technics:

> Every form of life, as Patrick Geddes has expressed it, is marked not merely by adjustment to the environment, but by insurgence against the environment: it is both creature and creator, both the victim of fate and the master of destiny: it lives no less by domination than by acceptance. In man this insurgence reaches its apex, and manifests itself most completely, perhaps, in the arts, where dream and actuality, the imagination and its limiting conditions, the ideal and the means, are fused together in the dynamic act of expression and in the resultant body that is expressed. As a being with a social heritage, man belongs to a world that includes the past and the future, in which he can by his selective efforts create passages and ends not derived from the immediate situation, and alter the blind direction of the senseless forces that surround him (319).

Human insurgence or agency, once understood through the self-invention and use of technologies to shape an immediate environment, could no longer be understood in a straightforward manner in modern times. At issue were not modern technics themselves, but the increasing conceptual distance between humans and the "extensions of their own organisms," the end products of industry, and a lack of awareness of the dangers and limitations to their uses (321). No longer were tools and utensils clearly and directly man-made, but they seemed to have "an independent existence apart from their user" due to their independent power sources and seemingly autonomous operation (322). It led to a focus on the products of the machine instead of the thinking that went into developing them. In fact, Mumford warned that in neglecting the "spirit" that went into the design, most products of technics were appraised solely as a means to an end. This displacement of agency in turn fueled the view that technics itself had become a creative force that was ordering a new environment and furthering a new mode of living. Mumford was quick to point out that these new modes of living were definitely man-made. As he stated, the new cultural values, and not the constantly outmoded instruments or quickly consumed goods, were the most lasting accomplishments of the new technics. These mainly beneficial contributions of the machine to culture included the "technique of cooperative thought and action . . ., the esthetic excellence of the machine forms, the delicate logic of materials and forces" (324). As a collective whole, he embedded these technics into culture by designating them as "the machine canon" of the arts. He insisted that through these values, humans had created an "independent environment that has reacted upon" other parts of culture and society.

The machine, when understood in Mumfordian terms, is as an aspect of culture from which norms are created. Society, he wrote, had been affected in the following ways: through a neutrality of order that was set neither by gods nor monarchs, but said to be increasing by "power, economy, objectivity, the collective will," through the new aesthetic experience that had been created by the machine and interpreted and assimilated by artists, especially in photography and in films,[144] through a growth in the importance of function that had been effected by countless engineers and machine workers,[145] and through a massive complication of the environment (330). The result of these various strands is what Mumford called the "objective personality" (359). And he took issue with what he felt was the recent dominance of this side of the human character.

Through his writing, Mumford hoped to do some of the preliminary work needed to change the focus of society toward more human and organic development. One step was to clarify the imbalance in the human personality to his readers. He felt that the effects of a mechanical environment had gone unnoticed by those that "live in this medium, and who have been formed by it, who constantly breathe it and adapt . . . to it" (360). Another step involved not only more awareness of the current mechanical environment, but also a fundamental critique of the dominant economic system. While mindful of the symbiotic relationship of technics and war, Mumford once again had an eye on fostering human development. He sympathized with pacifist and egalitarian ideological positions, arguing for a need to stop war and fundamentally reform capitalism to ultimately go beyond the machine and have more collective benefits from developments in technology: "Capitalism, which along with war played such a stimulating part in the development of technics, now remains with war the chief obstacle toward its further improvement" (353). In a series of quotes, he suggested that machines were inherently communist and that the economy should be so as well:

---

144 "Through the machine, we have new possibilities of understanding the world we have helped to create" (*Technics and Civilization* 343).

145 "Expression through the machine implies the recognition of relatively new esthetic terms: precision, calculation, flawlessness, simplicity, economy" (*Technics and Civilization* 350).

> Our modern technology has, in its inner organization, produced a collective economy and its typical products are collective products. Whatever the politics of a country may be, the machine is a communist: hence the deep contradictions and conflicts that have kept on developing in machine industry since the end of the eighteenth century. At every stage in technics, the work represents a collaboration of workers, themselves utilizing a large and ramifying technological heritage (354).

The problem was that this radical view of the machine had been concealed by the form of economy and its medium of exchange:

> In money-ridden societies, where men play with poker chips instead of with economic and esthetic realities, every attempt is made to disguise the fact that the machine has achieved potentially a new collective economy, in which the possession of goods is a meaningless distinction, since the machine can produce all our essential goods in unparalleled qualities, falling on the just and the unjust, the foolish and the wise, like the rain itself (355).

Before the dawn of what he envisioned as being a "Biotechnic phase," Mumford stressed the need to cultivate and assimilate the machine arts before it was too late to respond to human needs. In this last step, it was vital for society to "accept the practical benefits of the machine," but also to embrace its "moral imperatives and its esthetic forms" (355). This corrective involved further assimilation of the machine, machine principles and their integration to form a new way of living in a mechanical environment: "The economic: the objective: the collective: and finally the integration of these principles in a new conception of the organic – these are the marks, already discernible, of our assimilation of the machine not merely as an instrument of practical action but as a valuable mode of life" (356).

The final chapter of the book offers Mumford's overly optimistic assessment of an imminent future and swift claim that the "forces on the side of life" were beginning to grow from within technics (368). In retrospect, these observations seem rather out of step with historical events at the time. For example, he foresaw Western civilization as having entered a phase of "disassociation between capitalism and technics" just as Nazi Germany, which he visited, had developed not only a mechanical marriage between authoritarian technics and industrial capital, but also an anti-life manifesto with a truly rotten core: a racist worldview which could serve as the polar opposite of Mumford's organic and inclusive, utopian future (366). While in academic isolation writing *Technics and Civilization*, he underestimated the insurgence and repression of Hitler and the Nazi Party and overestimated the power of civil society and the arts to sustain democracy. In an attempt to connect the macro framework of his periodization with a radical ideology of history, Mumford seemingly neglected the insurgency that occurred instead, so quickly and intensely, on the side of the machine itself.

To his credit, Mumford's conscious use of the "prophetic form" throughout the last chapter, intended to compel his readers to action, represented a creative attempt to address needs at that time (434). What were some of these needs and did Mumford actually recognize negative potential for the future in the supposed goal of technics? He wrote that, in the search for regularity, order, and certainty that were the ultimate reasons for developing the machine, the machine had become "a goal of desire" (365). Now that society had become mechanically conditioned, his warnings against a sinister motive in technics went underrecognized. Mumford cautioned that aspirations themselves had become more machine-like and power-oriented, "the mechanical discipline and many of the primary inventions themselves were the result of deliberate effort to achieve a mechanical way of life: the motive in back of this was not technical efficiency but holiness, or power over other men" (364).

Instead of a teleology of power, Mumford offered a radical ideology, laid out in the last chapter, which favored egalitarianism through general economic, social, and cultural policies. Calling first for "changes in our conceptual apparatus" that are paralleled by "changes in personal habits and social institutions," Mumford sought a more organic and cooperative future (370). The over-importance of warfare, mining and finance should be replaced by an application of modern technics that placed increasing emphasis on harmony, balance, and equilibrium. The "failure to frame a comprehensive scheme of ends" in ways other than through work and consumption should be replaced by the clear goals of achieving "leisure, free activity, creation, . . . health, and biological activity, and esthetic pleasure" by radical means (378–379, 398). These included but were not limited to: increasing conversion, economizing production, normalizing consumption, implementing basic communism, socializing creation,[146] freeing workers from servile work or slavery, favoring amateurism as a corrective to automation and machine standardization, decreasing the machine, promoting a dynamic equilibrium in the environment, in industry, in agriculture and in population, but most radically taking political control of industry, of standards and quality of consumer products and of services like public banking, and by taking ownership of "land, capital, credit, and machines" (421).

146 Mumford goes as far as to label creative activity the "only important business of mankind" (*Technics and Civilization* 410).

A Predictable Ending?

How should one make sense of an ideological ending to a history of technology? More precisely, what does the last chapter of *Technics and Civilization* demonstrate about Mumford's understanding of history as a dynamic and contingent process of social change. It may seem odd that an account of the machine in history should include an entire chapter on orientation. Yet, for over seventy pages, in the longest chapter in the book, Mumford wrote about the future. One reason for including the chapter is that he did not metaphysically separate the future from the past and the present. One category of time flows or crashes into the next, each only a part of a greater cycle of organic time. As an author-historian employing the organicist mode of explanation, he absolutely integrated the three times, past-present-future, into one historical process. His word for this process is civilization, a historical world-making constantly being formed by the interplay between the human mind and its environment and expressed over long stretches of time. The result is a reification of abstractions like technics, capitalism, and the machine into enduring historical phenomena. Similarly, using the contextualist mode of explanation, Mumford relatively integrated his account of the *present past* of technics into a larger context of social change, which included a future that he wanted to prevent from becoming overly mechanical. A second reason to include the chapter seems to reflect the radical ideology he had learned under the tutelage of Veblen's books and concepts. Mumford's choice to write about the practical past (meaning moral past) was part of his intention to envision a future that included "socialized monopolies of [renewable] energy,"[147] "conscious economic regionalism," and a "vital standard of consumption" in an attempt to change the norms that governed society (380, 388, 399).

Combining his organic understanding of time with a contextualist and radical understanding of history, Mumford closed the book on a literal high note by once again stressing the need to bring technics into harmony with the new societal and cultural values he had laid out in the book.[148] Mumford was clearly eager to see hu-

---

147 In a conceptual parallel to Etzler's *Paradise*, Mumford argues for increased conversion and planning of renewable resources. "To increase conversion, then, is no simple matter of merely mining coal or building more dynamos. It involves the social appropriation of natural resources, the replanning of agriculture and the maximum utilization of those regions in which kinetic energy in the form of sun, wind, and running water is abundantly available. The socialization of these sources of energy is a condition of their effective and purposive use" (*Technics and Civilization* 382).

148 Mumford ends the book using an orchestra metaphor to lay out the task of bringing technics in tune with human needs (*Technics and Civilization* 434–435).

mans take the initiative, but as a historian, he wisely separated his own desires from historical realities and refrained from making any predictions about the future:

> The belief in the greater dominance of mechanism has been reinforced by a vulgar error in statistical interpretation: the belief that curves generated by a past historic complex will continue without modification into the future. Not merely do the people who hold these views imply that society is immune to qualitative changes: they imply that it exhibits uniform direction, uniform motion, and even uniform acceleration-a fact which holds only for simple events in society and for very minor spans of time. The fact is that social predictions that are based upon past experience are always retrospective: they do not touch the real future (423).

A positive future, if Mumford had had his way, would have traced its path directly through the machine. But the humanizing of the machine would involve major changes in ownership and in his words, "naturally, this means revolution" (421).

Mumford and Technological Change

Mumford wrote one of the first modern accounts of the history of technology in English. By doing so, he directly engaged in a discourse on technological determinism and addressed the question of whether technology drives history. As an author-historian, he was open to various modes of representation and not entirely bound to the traditional, realist writing espoused by most scientific historiography. This meant that Mumford could choose which aspects of reality to emphasize. The metaphors that he employed in *Technics and Civilization* should in no way be confused with mechanistic understandings of historical causality. They are rhetorical devices intended to represent the *present past* of technology that he had envisioned.

Using a common metaphor at the time, he alternatively defined the machine as a mechanical mode of life derived over a long period of time from cultural practices. Throughout the book he contextualized the many factors that play a role in technological change. However, from a mechanistic perspective, the metaphor (life = machine) can be construed to reduce causality as in the phrase, the machine drives historical realities. This misinterprets Mumford's position and would imply that he argued that the machine makes society mechanical. Another example of a figure of speech was used to associate technics with the human personality. Given the influence of biology on Mumford's views, his use of metonymy should hardly be surprising. Just as the word suit can refer to a businessperson, Mumford used the word personality to refer to the organic whole of human behavior and its interactions and developments within an environment. In this case, the expression of traits is what comprises technics. It is easy to misinterpret his use of rhetorical devices as mechanistic causality, which produces linear thinking: personality affects form, technics drives civilization; but these caricatures of Mumford's ideas do not reflect how he thought actual technological change occurred.

Rather than heightening the salience of machines, Mumford sketched a past that cast humans and human culture in the main role of the technodrama. Indebted first to Patrick Geddes for giving him a healthy skepticism of systems, Mumford's account of the history of technology did not see agency solely in the products and practices of industry. As a young man, Mumford had also learned from a contemporary of Geddes that "no single factor or set of factors is more ultimate than others" (Cooley qtd. in Mumford, *Sketches from Life* 158).[149] Mumford preferred a past that went beyond deterministic concepts of technological change and included the ability for humans to be insurgent, or to "overcome the conditioning process":

> Geddes found in this Lamarckian notion[150] a comforting alternative to some of the more uncompromisingly mechanistic interpretations of Darwinism. Later, Mumford would turn Geddes's idea of insurgency into an argument for human control of technology . . ., insisting that while the machine has long dominated us, we have within ourselves the capacity to regain control of our lives and our futures (Miller, *Lewis Mumford: A Life* 70).

Building on these ideas, the technological change in *Technics and Civilization* can best be described as dialectical, an interplay between the internal mind and the external culture, between individuals and society, and between the conditioning and insurgency of organisms and their environment. In the book, this dynamic led to contextualized outcomes as he detailed in the three phases. But beyond the particular, Mumford also assessed the cumulative effects of technics in civilization, meaning the larger context of social change. In his mind, they added up to a pattern that the author wanted to urgently counteract. And he concluded his argument by emphasizing the need for humans to reorient and radically change technics.

Mumford was conscious of the historian's role in forming new conceptions and contextualizations of technics. This can be observed in the originality of the ideas in *Technics and Civilization* in the context of North American writers. At a

---

149 Mumford credits Charles Cooly's monograph *Social Organization* (1909) for not looking at just parts, but organic wholes in history.

150 Lamarck's theory – "inheritance of acquired characteristics through environmental adaption," which states that "organisms make efforts of their own to transform themselves as well as their environment" (Miller, *Lewis Mumford: A Life* 70). The "capacity of the organism to overcome the conditioning process" was called *insurgency* by Geddes, and according to him, "reached its apex in man."

time when others in his intellectual circle were advancing the notion that technology had direct influence on society, Mumford was able to look past the appearance of lawfulness and predictability in historical and technical processes. Embedding his historical account of technology into culture, he sought to induce a broader understanding of the contingency of technological change and the need to improve the greater human condition in a mechanical environment.

Mumford's later writings provide more context for his initial thoughts on the matter. In a 1961 article published in *Technology and Culture*, Mumford challenged historians of technology to maintain a wider focus and look for clues outside a narrow purview when investigating technological change. He cited a "distortion of evidence" in the history of technology up to at least 1960 due an "excessive preoccupation" with technology understood as a concrete object ("History: Neglected Clue to Technological Change" 230–232). What he was referring to was a historiographic tendency to judge cultures and societies to a large extent based on the products they had developed. The task of attributing causality in history, he maintained, should not neglect clues from history. Not the kind of history, as he pointed out, that just records events as being caused by technology, but rather a more general, integrated history that seeks to interpret "the changes and transformations of a whole culture" of which technology is an "essential component." This integrative approach to the history of technology, which conceives of technology as part of a larger, "dynamic social context," necessarily rejects a concept as defined by material products of cultures and societies. At the very least, Mumford warned that such a limitation strips an account of technological change of the greater historical context in which it emerges and its meaning in society. As he argued, technology, being a component of culture, changes and is changed by many "aspects of society."

Mumford insisted that technological change stemmed partly from the "inner development of technology" (232). This pertained to the changes to the human mind and culture that had preceded outward manifestations of technical brilliance, of which the building of the Great Pyramids was his prime example. The concept of technology that Mumford sought to expose using this description was the technics of power concentration and authority that enabled their construction in the first place. Following Bimber's norm-based account of technological determinism, once firmly entrenched, these social norms set the limits for and can cause historical change. Keenly aware of the logical consequences of his argument, Mumford clarified his argument by rejecting the idea that the internal development of technics had caused the construction of the Great Pyramids. Instead, he underscored the "magnification and exaltation of human power and purpose" that had provided the proximate means to complete the project. Suggesting that historians had been looking at the wrong evidence,[151] he argued against a strict interpretation of historical causality by machines or technology. In the case of ancient Egypt, he asserted that "human purposes determined technological processes." Yet the greater argument that Mumford was making ultimately attempted to embed the story of the end product (the Pyramids) into what can be called "sun-god technics," or the social conditions that had enabled the human machine to come into existence. Once certain social conditions were present, the human-implemented technics gave rise to the big human machine, which then effected the construction of the Pyramids. In this and other writings, the contingencies of social relations are a necessary part of the historiography of technology. When a historian chooses to ignore following what Mumford called "that clue," but could be translated as 'this fact,' meaning the co-appearance of "large-scale technical enterprises," "political absolutism," and "military compulsion," causality in history is falsely attributed solely to inventors or historical periods in which machines and technology emerge (233–234).

Mumford's writings indicate that he understood the urgent need to challenge the common sense around how technological change occurred. He was certainly not the first to do so in general, but surely one of the first historians of technology in North America of this type. As detailed in a later chapter of this book, by mid-century some historians had heeded his call. In a 1962 article that describes a historian's perception of a shift in emphasis from past to future, Mumford underscored the general trend towards a more contingent understanding of technological change. He wrote that the pivot from historical causality by strict determinism to one by "organic events," "a world where purpose often controls process," was first explored by nineteenth-century historian Henry Adams. The problem was that very few had taken Adams seriously and that his own (meaning Adams) narrow methodological framework of determinism and focus on minute empirical details had not allowed him to formulate a response to how large-scale cultural transformations could have taken place ("Apology" 203). Adams's inability to plot out a new direction himself had opened the door for Mumford to expound a historiographic methodology that could do so. It began with the rejection of a completed or strange past.[152] "The weakness of a retrospective conception of history" is that such an orientation "confines itself to a dead past" (201). It resolutely included authorial voice. On the authoring process of historiography, Mumford pointed out that when the observer (author) "who meticulously leaves himself out of the picture leaves out the most sensitive instrument available for registering distant social

151 He cites a lack of awareness of the machine composed of human parts that helped build the Great Pyramids ("History: Neglected Clue to Technological Change" 233).
152 These terms are explained in appendix 2 of this book.

tremors beyond the horizon of direct observation" (202). It also involved speculation and a different kind of interpretation. Referring to what he termed "stultifying requirements of historical interpretation," Mumford rejected the uncritical basis on which nineteenth-century historiography had rested, that being the "belief that precise observation and accurate description are the main requisites for achieving adequate truths" (201).

In a revealing section of his later two-volume history of technics, Mumford expounded on his understanding of temporality and the mechanism of technological change. He suggested what he termed "organic continuity" is an important mechanism of historical causality. No longer as radical as he was in the pages of *Technics and Civilization*, and no longer as interested in the exact trajectory of technics, the older Mumford chose to underscore purpose over other factors in questions about technological determinism. He differentiated between nineteenth-century historical training, which had equated reality with "serial time and causality," and the radical shift in historical understanding that more recently put emphasis not on the past, but the future. This "organic order of temporal duration, of phylogeny and social inheritance (memory and history), in which both intersect in the present" is the result of uncountable "reactions in a more complex field modified by the organism's inherited nature and by lifetime accumulations of experience in an environmental field teeming with other organisms" (*The Pentagon of Power* 235).

After looking back on the material he had covered in *Technics and Civilization*, Mumford was more aware of the "irrational factors" that had become part of "machine-oriented technics" (*The Pentagon of Power* Acknowledgements). This must have confirmed in him the importance of the historian's insurgent role in changing their environment, which also included a reframing of the past. According to a later piece that he authored, the responsibility of the historian is to "create an ideological fabric capable of doing justice to every dimension of experience: not alone to causality and chance and statistical probability, but also to historic continuity and organic purpose, to creative choice and conscious design" ("Apology" 204). In his opinion, the "vital task" of the historian is that of "interpretation, prognosis, and constructive proposal, particularly in re-examining institutions," like war, that threaten "advances in culture" (207).

One might say that Mumford thought understandings of technological change flowed best through his own capable mind, especially one that could communicate with such purpose. In a section of his notes entitled "Thirty Years Before McLuhan," Mumford had observed the increasing immediacy and global nature of communication seemingly due to new technologies, but did not think that increased contacts would alter one's personality to make it "less trivial or . . . less parochial" (*Findings and Keepings* 328–30). By using technologies just because they are available or convenient, the "danger common to all inventions" is revealed: “a tendency to use them whether or not the occasion demands.” Instead of the positive image of the “global village,” Mumford thought that an increase of immediate contact would more likely lead to an increase in “the areas of friction” between people. Viewed positively, Mumford could imagine new communication technologies as a potential benefit to society as long as humans could make “use of mechanical apparatus to duplicate and further organic operations,” instead of “displace[ing] the human being” and being able to “re-focus him and enlarge his capacities.” Though he thought the dangers posed by multiplying technologies like TV and radio seemed “greater than the benefits,” he did concede that “no satisfactory answer” could be given based on “technics alone.”

Chapter 5
In-Formation Technology: An Ungrounding Experience

Marshall McLuhan's Aesthetic – Epistemic – Material Relations to the Past

There is unanimity among historians in presenting their subjects as figures without grounds.[153]
Marshall McLuhan and Barrington Nevitt, "The Argument: Causality in the Electric World"

What a man is saying is far from the effect he may wish to have, or that, in fact, he does have. Personally, I consider the effect that a writer wishes to achieve as his theory of communication.[154]
Marshall McLuhan, letter to Ashley Montagu

Umpire of Communication?

The previous chapter made the argument that Lewis Mumford's *Technics and Civilization* represents a major shift in English-language accounts of technology written by non-professional historians. If this claim is accurate, after the book's publication, scholarly accounts of the technological past could no longer be solely narrated to trace human progress through the ability to invent and innovatively use tools. As an author-historian, Mumford was able to historicize the notion of technology in the book as being one part of a broadly human-driven enculturation process. This was significant because *Technics and Civilization* was written for an English-speaking audience steeped in the suddenness and materiality of what historians refer to as the Second Industrial Revolution. In counteracting this common narrative, Mumford's account made it possible to view technological development as the

153 One of McLuhan's main critiques of Western scholarly practices was a tendency to focus on proximate causes separate from the ultimate environments conditions which are the grounds for a certain human behavior or cultural practice to emerge.
154 A second critique follows from this main critique and concerns accepted ideas surrounding the notions of technological and historical change. McLuhan frequently cited the "sensory bias" of "literate visual modes of perception" as, among other things, contributing to a widening gap between cause and effect (McLuhan and Nevitt 3–4). In a letter to anthropologist Ashley Montagu, McLuhan lamented the lack of scholarly interest in the "effects of innovations." Given that "technologies create environments which alter forms of human perception," studying "non-lineal, non-sequential causality" may contribute to a better understanding of how media work (E. McLuhan and M. McLuhan VIII).

https://doi.org/10.1515/9783111396477-006

long-term and ongoing manifestation of the mind. It also maintained a critical space at the heart of any future analysis of a history of technology to be within a framework of human culture.

Beginning in the early twentieth century, and particularly after World War II, the emerging military, technological, and cultural hegemony of the United States forced people around the world to deal with a new colonial and global power. In an academic reaction from the edges of that empire, Marshall McLuhan wrote a number of books and articles attempting to connect technologies to their supposed effects in society. The Canadian scholar was drawn to reassess nineteenth-century notions of historical understanding that accompanied the term technology. Like his U.S.-American contemporary Lewis Mumford, McLuhan understood the modern world to be out of balance. He too saw a need to reconsider ideas of historical change to better understand how society was becoming less in tune with his own traditional values. McLuhan was well aware of the dominating cultural and linguistic influence emanating from the United States, and he sought to understand its workings first within the context of what he defined as literary texts – advertisements – and then as part of a much larger framework of human communication past and present – media.

Mumford and McLuhan never had a working relationship. Their sustained common interest in technology's effects in society resulted in almost no personal correspondence. Any notion of "exchange" between the two scholars was achieved almost exclusively by reading and responding to each other's work. Initially, McLuhan was able to regard Mumford as an ally in his own investigation of technology. Early in his career, McLuhan read Mumford's books, including *Technics and Civilization*, which he heavily annotated. He also wrote a review of Mumford's *The Culture of Cities* and had it published in a St. Louis University Student Literary Magazine in December of 1938. The extent to which McLuhan had gained theoretical insights from reading Mumford's work has already been established.[155] As divulged in a quote below, even McLuhan himself acknowledged having profited from reading Mumford's early books. But any characterization of McLuhan's admiration for Mumford must be balanced with a different understanding of technology and idiosyncratic relations to the technological past that eventually emerge from McLuhan's work. Skipping ahead many years, after the publication of *Understanding Media* in

155 Scholar James Carey points out that some of the arguments that McLuhan later became known for were foreshadowed by Mumford; just to name a few, "a reassertion of the organic principle," "emphasis on technology as extensions of man," "the effect of printing was to unbalance the human sensorium," and "the content of a medium being another medium," a notion which can be traced back even further to Mumford's use of Oswald Spengler's term "*Cultural Pseudomorph*" (Carey 173–75).

1964 had made McLuhan more widely recognized, Mumford also read and annotated McLuhan's famous book. Their subsequent writings clearly demonstrate that a major point of divergence was the way each scholar conceptualized technology's past.

Mumford considered McLuhan's ideas about electronic technology to be one-sided, unoriginal, and at worst, exploitative. Some responses to McLuhan's work can be extracted from chapter ten of his final major work, *The Pentagon of Power*, in which Mumford described "The New Megamachine" (*The Pentagon of Power* 293–99). Mumford was not impressed by McLuhan and his followers. Most references to McLuhan show a rather negative predisposition. In one particular section, entitled "Electronic Entropy," Mumford complained that McLuhan had framed the role of electronic technologies in society in too novel and optimistic a manner. Referring back to his own text in *Technics and Civilization*, Mumford reminded his readers that, in any case, assessing the potentialities of electronics was nothing new. He continued his overall rebuke of McLuhan by pointing out that he (Mumford) had also "anticipated its drawbacks," and that he had done it thirty years before McLuhan was considered by some to have been "the unique prophet of the Electronic Age" (*The Pentagon of Power* 295–96). Mumford's critical assessment of McLuhan and other "minds content to exploit the medium and ignore the message" can be summed up with his warning that electronic technology enables both "instant power" and "instant knowledge," but also "instant destruction." Critically, instead of perceiving electronic technology as an entirely unique innovation, Mumford preferred to integrate it into a continuing story of technics for which he "demand[ed] a more humanly valuable result" (299, 304).

McLuhan's work narrated a different story about technology. The Canadian understood the proliferation of electronic technologies to mark a significant break with the past, a change in dominant forms of human communication which he believed would necessitate an understanding and adjustment in cultural awareness. While it may seem obvious, historicizing technology in this way allowed McLuhan to make sense of what he perceived as major changes to society since the advent of electricity. Ultimately, it helped justify his stance on education in the second half of the twentieth century. Less clear is how McLuhan might articulate an argument that would differentiate and justify his approach to technology when compared to Mumford's *longue durée* approach to the history of technics and the development of human society. The aforementioned published attacks on his ideas had gone seemingly unanswered by McLuhan, at least formally and publicly. Scholars have been left to speculate as to McLuhan's critique of Mumford until a recent finding of an academic document in a box of unrelated material. In an unpublished, typed manuscript dated 28 February 1973, McLuhan's praise of Mumford's early writings, surfaces for the first time. In the manuscript, McLuhan even wrote that he had "incurred a debt"[156] from the American, and then proceeded to respond to criticism of his ideas that Mumford expressed in *The Pentagon of Power* (1970).[157]

Liberation from Sensory and Perceptual Effects of Media

In what follows, this new material is advanced to provide specific evidence of the different manner in which McLuhan approached the subject of technology. Not bound to German discourses on *Kultur* and *Technik*, McLuhan focused mainly on modes and forms of communication technology in history as grounds for investigation. At the beginning of his career, he wrote about technology in response to the mechanization of society. But after the publication of his first monograph, *The Mechanical Bride*, in 1951, he was forced to acknowledge that the effect of the dominant medium of communication, TV, and of other electronic media, had resulted in a change in his assumptions about society.[158] His subsequent work was no longer concerned with the process of mechanization, but became more focused on how technologies "actually affected their makers and users in their perceptual lives" and on developing the intellectual tools to deal with these accelerated effects in learning and on individuals ("Mumford Manuscript" 16). In this way, the short manuscript can be read as a justification for his critique of technology not only as a product of human culture, but also as a shaper and constrainer of human sensibilities and awareness.

McLuhan's ideas concerning technology are also detailed against the backdrop of scholarly historiography that had begun to shift. They are distinct from notions that respond to the mechanization of society and can be seen as coinciding with ongoing discourses on emancipation of the individual from various forms of constraint, which gave a new perspective to critical thought about the technological past. McLuhan's contribution to the existing scholarship on the history of technology was that he wrote about the subject in conceptually different ways and understood the human past not only as a process of mechanization, but

156 The same phrase he used in writing I.A. Richards about his influence (*Letters of Marshall McLuhan* 355).

157 Reference to McLuhan's "debt" is quoted from page one of an unpublished manuscript that will be referred to as the "Mumford Manuscript" for the remainder of the text.

158 In an interview in the June 1967 issue of *Encounter*, McLuhan admits to a reversal in his thinking about the behavioral patterns of society as a result of a new dominant medium: "*Mechanical Bride* is a good example of a book that was completely negated by TV. The whole mechanical assumptions of American life have been shifted since TV; it has become an organic culture" (McLuhan and Stearn 50).

as one that can be divided up into various epochs of media dominance that affected networks of communication. He attempted to find the right intellectual tools to educate about the importance of language, awareness, and perception to complement a pursuit of "mere sciousness," or solely knowledge ("Mumford Manuscript" 4). His unorthodox historiographical style, his almost singular focus on form as opposed to content and his attempt to revive an obscure type of causality, all represent examples of a larger break from modern historiographical traditions in his approach to writing history, in the presentation of his ideas and in the challenge to accepted notions of reason, science, truth, and progress.[159] If a historian is understood to be a judge, his approach coupled with the context of changes in historiography render McLuhan rather atypical of a historian in the modern sense. He should be characterized not as a representational arbiter of facts about the mechanization of society or an objective umpire of communication, but more as a literary scholar who advised students to strive to understand the medial past in order to "create an available means of liberation" from the dominant media environment of the present ("Mumford Manuscript" 10).

Back to the Poetic Roots

McLuhan's attempt to understand technology and the effects of the resulting changes to sensory ratios over time has led to a thematic association with communication and media studies. One side-effect is that his analysis is more frequently judged by scientific and realist criteria.[160] Both of these tendencies conceal a more appropriate context to understand McLuhan's writings. As much as he favored multidisciplinary learning, the reader is advised to keep in mind that he was instructed in the methods of literary studies and became a professor of literature. As historian James Curtis once mused, McLuhan being characterized as a media theorist disconnects him from his theoretic sources.[161] It also does not ac-

159 Here is meant the antinomies or tensions between supposed modern and postmodern historiography mentioned in chapter one of this book, namely (1) objective vs. normative (2) found vs. invented (3) factual vs. fictional (4) representational vs. perspectival (5) social vs. cultural (6) real vs. textual (7) explanatory vs. interpretive (8) deductive vs. inductive (9) macro vs. micro (10) necessary vs. contingent (Susen 136–170).
160 In this instance, scientific means natural science and realist alludes to the literary category of realism.
161 Curtis argues that "interpretations of McLuhan arising from this classification [as a media theorist] are usually simplistic because they dissociate McLuhan's theories about media from the philosophy of history which informs them" (144). Curtis traces McLuhan's ideas back through European intellectual history in which "analogous assumptions" were made by Hegel and Bergson.

count for his academic training under I.A. Richards in the symbolic and emotive functions of language. The result of such oversights may lead to an undiscerning focus on McLuhan's more provocatively formulated statements, separated from their disciplinary context. Considered solely in their finalized and published book versions, any determinism associated with McLuhan has been used to strip his ideas of the contingent framework in which he regarded technological change to operate. McLuhan was not your typical professional academic historian writing in the realist mode. Even a close and careful reading of his scholarly writings can obscure his interpretation of the extent to which technological, historical, and cultural changes can affect individuals and, in turn, greater society. This is why, before examining selected McLuhan writings in a close reading, several aspects of his non-traditional approach to historiography need to be articulated and reinforced. These include, specifically, scholarly influence on his thought, method, and sources by I.A. Richards and Harold Innis. Taken together, they ultimately hint at McLuhan's particular relations to the past and help more accurately represent an answer to how he might respond to the question: does technology drive history?

I.A. Richards, Training in Perception, Poetic Underpinnings

McLuhan's time as an undergraduate student and young teacher not only confronted him with various literary ideas and authors, but also with three different cultural environments. Spanning ten years, from 1932 to 1942, McLuhan studied at the University of Manitoba (B.A. 1932; M.A. 1934) in Canada and pursued a second set of studies at Cambridge University (B.A. 1936; M.A. 1939; Ph.D. 1942) in England. He went on to teach at the University of Wisconsin-Madison from 1936 to 1937 and at St. Louis University from 1937 to 1944 in the United States. On a personal level, these changes in scenery gave the young McLuhan an opportunity to leave his home and to observe cultural differences. The early experiences in different coun-

Hegel states that history should not be understood as an individual process, but rather as a "supra-personal" one and the interpretation that an externality is "the projection of the form of the organism into the external world" (147). The author credits Bergson with the notion that such extensions have psychological consequences. As a result of connecting McLuhan to Bergson, it is indicated that the logic of McLuhan's arguments can be seen as deterministic. As Curtis points out, "If people create extensions of themselves in tools, and if such acts of creation have psychic consequences, then we may find in Bergson the origin of what some have called 'technological determinism' in McLuhan" (148).

tries can be considered an informal training in perception. Formal instruction in the methods of perception occurred while McLuhan was in England.

The investigations that were later transformed into books and articles had their sources in the disciplinary training he had received at Cambridge University. During his PhD studies, McLuhan was taught by I.A. Richards, a professor and literary critic whose writings were influential to the formation and development of New Criticism. Among Richards' more significant books was *The Meaning of Meaning,* published in 1923, which investigated the influence of language on thought. In this study of symbolism, Richards and co-author C.K. Ogden called the meaning of words into question by analyzing the relation between symbols and referents in a context or "symbol situation" (Richards and Ogden 25). The authors argued that once the magic "power of words" is acknowledged as the "most conservative force" in life, a "contextual theory of signs" becomes vital in any study of meaning (47). What the book showed a young McLuhan was the futility of defining complex things merely by identifying their referents. One had to look at experience and surroundings to understand meaning. The experience in the first part of the twentieth century was that the growth of (scientific) knowledge had not assuaged feelings of uncertainty in the meaning of everyday life and of everyday things. To understand meaning in a technological and mechanized society, McLuhan leaned on his training of perception in literary studies. Any investigation of the present relationship between technology and society would have to be conducted through an interpretation of text through language as his teacher had laid out: interpretations of signs begins with a "psychological reaction to it determined by our past experience in similar situations, and by our present experience" (Richards and Ogden 244, 246).

I.A. Richards was not only interested in the symbolic functions of language, but also in the emotive functions and their consequences for human experience. In his 1926 book *Science and Poetry,* one can see many methodological themes that will reappear in McLuhan's later work. First, Richards spoke to the effect of form on content: "In nearly all poetry the sound and feel of the words, what is often called the form of the poem in opposition to its content, get to work first, and the sense in which the words are taken is subtly influenced by this fact" (23). Second, Richards was interested in the value of experiencing poetry from beginning to end. In discussing the place and future of literature, Richards asked, "what is poetry, how can it be used, and is it valuable?" (9). In order to analyze the experience of reading poetry, he advanced metaphorically "from the surface inwards" or from the physicality of a written text: "The surface is the impression of the printed words on the retina. This sets up an agitation which we must follow as it goes deeper and deeper" (11). This inductive line of inquiry was later used by McLuhan to delve into what associations with the term technology could mean and which consequences it may have for perception. Finally, Richards' influence on McLuhan's thought also extends to the importance of equilibrium, order, and coherence. The English scholar saw the mind as a system of balances, which are one's interests,[162] and posited a poetic relation to the world as valuable: "poetic experience as the swinging back into equilibrium of these disturbed interests" (15, 18–19). To report on his later research on forms of media, McLuhan must have taken some clues from the differences Richards described between a scientist and a poet. As Richards stated, "it is never what a poem *says* which matters, but what it *is*. The poet is not writing as a scientist. He uses these words because the interests which the situation calls into play combine to bring them, just in this form, into his consciousness *as a means of ordering, controlling and consolidating* the whole experience" (25–26). If, as Richards suggested, "the chief characteristic of poets is their amazing command of words," then McLuhan might have been convinced to look to these "the master[s] of speech" to also be "master[s] of experience" (38, 41). Such conflation of expertise in language and experience was ultimately thought to be justified because McLuhan, like his teacher Richards, considered these master perceivers to be best suited to make sense of experience because "the business of the poet . . . is to give order and coherence" (Richards 55).

With respect to understanding McLuhan's academic background, a few more teachings of Richards are worth mentioning. Richards emphasized fictional representations to counter overly factual ones. To provide more evidence in his argument for the importance of poetic approaches, Richards explained the difference between a pseudo-statement in poetry and a statement in science: "a pseudo-statement is a form of words which is justified entirely by its effect in releasing or organizing our impulses and attitudes . . . a statement, on the other hand, is justified by its truth, i.e. its correspondence, in a highly technical sense, with the fact to which it points" (58–59). Aware of the potential to misunderstand the message of poetic wording, he lamented its confusing relationship to the truth: "A pseudo-statement is 'true' if it suits and serves some attitude or links together attitudes which on other grounds are desirable. This kind of truth is so opposed to scientific 'truth' that it is a pity to use so similar a word." With this distinction in mind, McLuhan could later draw on the use of a pseudo-statement to order emotions and attitudes, something for which Richards had laid the groundwork. In a statement that unknowingly foreshadows McLuhan's first book, Richards, referring to poetry, asserts that "it is capable of saving us; it is a perfectly possible means of overcoming chaos" (82–83). In fact, the manuscript that eventually be-

162 Richards belief that "man . . . is a system of interests" is reflected in McLuhan's later focus on the human sensorium (Richards 21).

came *The Mechanical Bride* (1951), was given the working title *A Guide to Chaos*, and is an endeavor to juxtapose what McLuhan considered to be poetic texts – advertisements, comics, and newspaper clippings – with his own analysis. McLuhan's subsequent direction was influenced by a Canadian colleague. Yet, in what can be read as a retroactive challenge to McLuhan, Richards appeals for thorough investigation of the truth of a matter on poetic terms: "attempts to use poetry as a denial or as a corrective of science are very common, one point can be made against them all: they are never worked out in detail" (63). The detail of a poetic investigation into a "scientific" topic would be worked out after this first book. More than ten years passed between *The Mechanical Bride* and McLuhan's two other major works *The Gutenberg Galaxy* and *Understanding Media*.

Quantum Leap in an Understanding of Technology's Past

During the late 1940s, McLuhan's focus began to shift from purely literary studies to an interest in technology and communication. He later disclosed in an interview that the switch did not represent a total break from his previous research, but rather an expansion of it (McLuhan and Stern 50). What prompted the transition to a seemingly unrelated and new subject matter? Besides reading Mumford, McLuhan had greatly profited from art historian Sigfried Giedion's books *Space, Time & Architecture* (1941) and *Mechanization Takes Command* (1948). McLuhan considered Giedion to be the "first technological student to give adequate and particularized attention to the changes in human sensibility relating to the production and use of everyday objects" ("Mumford Manuscript" 16). McLuhan's subsequent scholarship on technology thus pursued the line of research initiated by Giedion, who had studied "the development of machines and industry as art forms" (McLuhan, "Guaranteed Income in the Electronic Age" 75–76). McLuhan had also moved back to Canada to teach at the University of Toronto.[163] This is where, in a setting of Canadian politics and history, McLuhan's ideas about the relationship between technology and society, as Menahem Blondheim and Rita Watson observe, took a "quantum leap after meeting Harold Innis in 1949" (7). By the early 1950s, McLuhan had landed an article in the CIIA's *International Journal* that argued for the investigation into forms of communication and their impact

163 In 1944, McLuhan returned to Canada to teach first at Assumption University in Windsor, Ontario (1944–1946), and then at St. Michael's College at the University of Toronto in 1946, where he stayed for the next 30 years.

on culture and politics.[164] He later received funding from the Ford Foundation for a seminar on culture and communication.

Although any of the above could be viewed as a contributing factor to McLuhan's academic path, what is currently being suggested as part of a shift in discourses surrounding technology is particular to the connection between Innis and McLuhan: the context of American imperialism and the extracting of "communication" as a subject of scholarly inquiry from other disciplines. According to the aforementioned scholars, the two Canadians came to view communication as the "key to understanding the co-development of mind, culture and society" (Watson and Blondheim 7). The operative word in the last quote is co-development because it is an interpretation somewhat at odds with the one casting McLuhan as a technological determinist. Innis's investigation into the relationship between technology and society was foundational for McLuhan. And it is the adoption of Innis's framework that enabled McLuhan to take a leap forward in his particular understanding of the technological past.

Harold Innis, Media as an Ordering Concept for Time, Interface Methodology

McLuhan has often been quoted describing an uneven relationship between his and Innis's theoretical analyses of technology and society. The specific quote, which refers to McLuhan's second book, *The Gutenberg Galaxy*, considers the monograph to be a footnote to Innis's observations of the psychic and social consequences of writing and printing.[165] Instead of questioning the accuracy of this comparison, at a minimum it can be taken as evidence that McLuhan viewed his own work as an extension of Innis's. Just as a footnote extends its related passage,

---

164 The Canadian Institute of International Affairs (CIIA) and its successor, the Canadian International Council (CIC), are responsible for the publication of the *International Journal*. Despite the words international, affairs and journal in their titles, both the organization and the journal represent inherently political and national interests. According to their website, the CIIA was founded in response to "the experience of World War I and growing Canadian assertiveness and independence on foreign affairs" and has maintained a "constructive relationship" with the Canadian government. The academic journal was started in 1946 "to provide a platform and outlet for informed Canadian views on international affairs" (See entry "Our History" in the works cited section at the back of the book).

165 "I am pleased to think of my own book *The Gutenberg Galaxy* as a footnote to the observations of Innis on the subject of the psychic and social consequences, first of writing and then of printing" (McLuhan qtd in Innis ix).

so too did McLuhan attempt to provide more concrete, individual and detailed effects for Innis's larger framework of investigation.

Harold Innis (1894–1952), whose formal instruction at McMaster University in Toronto and at the University of Chicago was sandwiched around military service on the Western Front, had been trained in economics. After writing his PhD on the history of the Canadian Pacific Railway, he eventually became known for the Staples Theory of Economic Development. Yet later in his career he switched from studying staples (fur, lumber, wheat, coal, etc.) to studying media. This abstraction was described by McLuhan as Innis's attention moving from the "trade-routes of the external world" to the "trade-routes of the mind" ("The Later Innis" 385). In his final book, *Changing Concepts of Time* (1952), Innis himself provides a succinct summary of his later research:

> I have attempted elsewhere to develop the thesis that civilization has been dominated at different stages by various media of communication such as clay, papyrus, parchment, and paper [. . .]. Each medium has its significance for the type of script, and in turn for the type of monopoly of knowledge which will be built and which will destroy the conditions suited to creative thought and be displaced by a new medium with its particular type of monopoly of knowledge (78).

The best source for Innis's research on media forms and their effects in society is a monograph that was published with the title *The Bias of Communication* (1951).[166] The book is a collection of papers that he had revised to support the ideas from the preceding book, *Empire and Communication. The Bias of Communication* reflects on the question "why do we attend to the things to which we attend." The more provocative part of this thesis provides that changes in communication are the result of "the things to which we attend" (*The Bias of Communication* "Preface"). Innis begins the chapter from which the book takes its name by noting that "a medium of communication has an important influence on the dissemination of knowledge over space and over time and it becomes necessary to study its characteristics in order to appraise its influence in its cultural setting" (33). This recognition leads Innis to question the term history itself and the ability of the field of knowledge he was trained in (economic history) to overcome the limitations of what he calls "present-mindedness," "antiquarianism," "stagnation" and "maturity":

> History in the modern sense is about four centuries old but the word has taken on meanings which are apt to check a concern with facts other than those of immediate interest and its content is apt to reflect an interest in immediate facts such as is suggested by the words "all history proves." As a result history tends to repeat itself but in the changing accents of the

166 The 1964 version of the book includes the introduction written by McLuhan.

> period in which it is written. History is threatened on the one hand by its obsession with the present and on the other by the charge of antiquarianism (61).

The two passages above suggest that Innis was searching for a way to both rethink and reframe history to provide a more critical understanding of knowledge production. In the following chapter, Innis further develops his argument. "A Plea for Time," which could be retitled reflections on temporality and the reorienting of history ("the things to which we attend"), Innis suggests a remedy to the present bias of communication by imploring historians to "appraise the character of the time concept" (62), which will allow them to "escape . . . from our obsession with the moment and . . . from our obsession with history" (90). The Canadian scholar's proximate goal is to have his colleagues explore concepts of time when writing their accounts while attending to the effects of the time bias on representation in history. But what Innis ultimately has in mind is stability between the "demands of time and space" that allow for human flourishing. He argues that "a stable society is dependent on an appreciation of a proper balance between the concepts of space and time," and that "the character of a medium of communication tends to create a bias in civilization favourable to an overemphasis on the time concept or on the space concept . . ." (64). Such a balance, of enough historical knowledge to recognize patterns, but not too much so as to stagnate, is achieved not though the dominance of one medium of communication, but rather an appreciation of the various culturally-imbued concepts of time and space. Thus "A Plea for Time" can also be interpreted as an appeal to attend to various dominant types of time as ordering concepts in historical accounts.[167]

In Canada during his lifetime Innis could bear witness to a gradual change in power: the transition of economic life dominated by one monopoly of power (British Empire) to hegemony by another (American Empire). As the war economies raged, his writings concentrated on describing what he perceived as the neglected factor in twentieth-century economic history – the media.[168] In effect, Innis wanted to make a "critical survey and report" on mechanization of thought and the "monopolies of knowledge" that have resulted from it (190). Writing at a time confronted with state fascism, he thought that the "conditions of freedom of thought" were further "in danger of being destroyed by science, technology, and the mechanization of knowledge" (190). Towards the end of the book, Innis quotes British po-

---

167 He spends the next pages detailing various typologies of time: astronomical time, social time, biological time, Egyptian, Greek, Roman time, Christian time, mechanical time, industrial time, business time, corporate time.

168 One of the chapters of his last book *Changing Concepts of Time* is entitled "Press, a Neglected Factor in the Economic History of the Twentieth Century."

litical scientist Graham Wallas in his concern of the "danger that knowledge was growing too vast for successful use in social judgement" (191)

Not long after McLuhan had read Innis's writings, in a journal article entitled "The Later Innis," McLuhan both praised and criticized his colleague. Characterizing his approach as a "panoramic survey" of "great range and value," McLuhan subsequently described Innis's contribution to the history of technology as follows, "more than anybody before him, [Innis] concentrated on the changing social attitudes to time and space as they were modified by the material media of communication" (387–389). In a later book, McLuhan credited Innis with being the first to detect the "*process* of change implicit in the *forms* of media technology,"[169] but in this article he commented on Innis's lack of preparedness to confront the "total social process which is human communication and or participation." Placing the blame on the disciplinary boundaries of Innis's training as an economic historian, McLuhan wrote that terms like "technology" or "economics" are limited when trying to describe this totality. As McLuhan argued, it left Innis in a poor position to "grapple with art forms" and move beyond classifying the specific effects of media as a "bias." Though they shared an interest in the study of communication theory and practice, it was McLuhan, or so he thought, who was best suited to exploit "the tools of artistic analysis" to understand the particularities of how different media work.

Critically, Innis's technique of breaking up history into media phases[170] and his practice of associating changes in media to changes in society[171] was taken on by McLuhan. Later, when McLuhan wrote an introduction to the 1964 second edition of the *Bias of Communication*, he focused more on insights into Innis's methodology. Not occupied "in providing accounts of the contents" of various disciplines, McLuhan described Innis's writings as interested in "mutual interactions" and emphasized the importance of Innis's concern "with the unique power of each form to alter the action of the other forms it encounters" ("Introduction" ix–x). He praised Innis's research for "directing attention to the bias or distorting power of the dominant imagery and technology of any culture" and thereby helping his readers learn

169 See McLuhan, *The Gutenberg Galaxy – The Making of Typographic Man* 50.

170 For example, "The radio introduced a new phase in the history of Western civilization by emphasizing centralization and the necessity of a concern with continuity" (*Bias of Communication* 60).

171 For example, "The revolt of the American colonies, division between north and south, and extension westward of the United States have been to an important extent the result of the spread of the printing industry" (*Bias of Communication* 77) or "The coincidence with the advent of radio of dictatorship in Russia, Germany, Italy, Great Britain, the United States or Canada is not accidental" (*Essays in Canadian Economic History* 127).

"how to understand cultures" (xi). According to McLuhan, more conventional historiographical methodology attempts to classify change over time into narrated perspectives, a time-consuming structuring process which can be done more or less successfully. Yet in the introduction to the book, McLuhan cited Innis's proclivity to organize historical data in another way. Especially later on in his career, Innis became first and foremost a "recognizer of patterns" (vii–viii). Such a historian employs the aesthetic "technique of symbolism" as an "interface" to understand historical causality as a "complex process of interaction." In contrast to the historiographical method of classification, McLuhan stressed the "instantaneous" historical understanding or "recognition" that resulted from Innis's examination of past situations. McLuhan further described Innis's use of the act of history writing as a scientific space to test the ways the "character of technology" shaped cultures.[172]

When considering the introduction to *Bias*, at times it is hard to distinguish where Innis's ideas end and McLuhan's begin. Perhaps this was even the desired effect. Clearly, McLuhan's interest was piqued by Innis's rethinking of the past and his investigations of how various dominant media had shaped knowledge production. McLuhan was also impressed with Innis's methodology, which appealed to him as a professor of literature steeped in perceptual training. In a short foreword entitled *The Toronto School of Communication Research*, Elihu Katz summarized the research interests of both scholars as follows. Innis emphasized "media's ultimate effects as change in social organization of societies and institutions in the long-run in response to shifting media ecologies" (Katz 4). Innis's research purview was reduced from the macro to the micro as McLuhan attempted to understand media "as vehicles/engines of a change in the mental processing of individuals in the long run as a result of unique technological attributes of the different media."

Before moving on to McLuhan's own methodology and historical sources, one last, important aspect of Innis's research needs to be observed, his framework of contingency for technological change. As Ronald Deibert has noted, throughout his career Innis was generally interested in the "dynamics of large-scale social and political change" (Deibert 29).[173] According to Deibert, Innis's writings can be characterized as containing "radical contingency" (36). In an article concerned with "meta-historical traits" in Innis's writing, Diebert demonstrated that the Ca-

172 McLuhan refers to Innis's use of history metaphorically as a "scientific laboratory, as a set of controlled conditions within which to study the life and nature of forms . . ." ("Introduction" x).
173 Exploring three underlying consistencies or "meta-theoretical traits" in Innis's writing, Deibert classifies the Canadian theorist as being an "ecological holist" (30). He describes the traits as historicism, a combination of material and ideal factors and his focus on the biases of space and time in the analysis of empires and civilizations.

nadian scholar "saw history as open-ended, contingent, and without overarching purpose" that does not privilege "material, technological, or cultural factors" in his historical analysis (39, 52). The author could thus place Innis's writings in line with Mumford and other early-twentieth-century scholars who rejected both strong social constructivist and materialist perspectives in favor of an integrated approach to explain historical processes (41). Diebert added that Innis's emphasis on "contingency and historicity" emerged when "the momentum of the social sciences was in the direction of modernization and teleology" (51). In fact, Deibert built on this finding to suggest a possible reading of *The Bias of Communication* as an "extended essay on how historical circumstances and the constraints and opportunities of local context take history in particular directions" (37).

Could this characterization of Innis's research as contingent be extended to McLuhan? One would think that, in a context of an emancipatory discourse, the answer would be yes. Especially within a Canadian context, contingency in history is a fundamental framework to provide a counternarrative to military, technological, and cultural domination by colonial powers. In his book on *Technology and the Canadian Mind*, Arthur Kroker described Innis as a "technological realist" and McLuhan as a "technological optimist." Might these labels be interpreted as the frameworks in which each scholar confronted the chances of Canadians determining their own future?

McLuhan's Methodology and Sources

Returning to the comparison made between Mumford and McLuhan at the outset of this chapter, it may be helpful to quickly recap what came before changes toward postmodern historiography and a shift in the discourse of technology, namely Mumford and the historical method he employed. It is important to point out that, in a sense, Mumford's method in *Technics and Civilization* followed established norms in the modern discipline of history. He primarily collected evidence by researching archived documents and by tracing and compiling the historical causes of mechanical society. He then used what he had discovered or found to create a factual representation and to justify arguments about real, social developments of the technological past. The attempt to situate his findings in an opportunistic present moment also fits well within a certain historiographical mode and tradition. What seems most ahistorical about the book is that he concluded the account projecting a utopian social order that would help create the basis to integrate the mechanical into the organic.[174] But like any conventional historian, as time passed, Mumford looked at new and different sources and eventually realized that themes from his first book on technology would need to be expanded and revised because, as James Carey has pointed out, "by mid-century he [Mumford] could see no difference between the capitalist and socialist state, as both were dedicated to an extirpation of the past, total management of the present, and a future based solely on the mechanics of power and productivity" (176).

In his final two volumes on the history of technology, Mumford's method combined what he termed "disciplined speculation" with literary acumen to produce an even longer monographic account.[175] Mumford was certainly novel in having produced generalist and readable histories of technology vis-à-vis the mechanization of the human mind. And by admitting non-material evidence to what is considered scientific, Mumford achieved his own goal of rooting culture in an understanding of the past by attempting a fresh look at the causes of the (then) current situation to explain its effects. Despite exhibiting a rich historical imagination in his writings, his approach does not represent a clean break with disciplinary conventions. He was however successful in expanding the historical method beyond conventional sources and methods as well as openly embracing the connections between technics and culture.

Privately, McLuhan once critiqued this approach to understanding technology and culture, referring to the American scholar as "Mumford the urbanite," who argued in form of an "abstract assertion" for a "renewal of the human condition" (*Letters of Marshall McLuhan* 183–184). To guide his own investigation of technology, McLuhan was uninterested in formulating factual statements based solely on research of historical sources. Instead, he also preferred the "cultivation of positive *grammatica*" to enlist the insights gained by poets. Mumford's theater metaphor and emphasis on the city and an unfolding of the technodrama was replaced with a medical metaphor and emphasis on human sensibilities (experience). His main motivation for this was divulged much later when McLuhan mentioned that he had shelved the idea of assessing technology from a fixed viewpoint with a specific aim, à la Mumford, on the grounds that it was too ambiguous.[176] Focusing in on the

174 Among other sections in his final chapter entitled "Orientation," Mumford argues for "Basic Communism." He is careful to avoid associations with Marx while conceding that, "the classic name for such a universal system of distributing the essential means of life . . . is communism . . ." (*Technics and Civilization* 400–406).

175 See chapter two of *The Myth of the Machine* (1967), where Mumford argues for the "need for disciplined speculation" about the technological past (14–15).

176 The following quote was taken from what, in this book, has been referred to as the "Mumford Manuscript." Gordon's use of the quote in his book alerted me to the manuscript's existence.

category of experience, McLuhan became interested in how profoundly people could be affected by their technical-cultural environment. And because he thought that no amount of knowledge could be acquired to protect oneself from novel sensual experiences and patterns engendered through new technologies, he attempted to examine its effects in order to "acquire the antidote."[177] If, as he once claimed, technology "tend[s] to produce artifactual environments," then he might find the corrective by going back to the aesthetic roots and enlisting the help of artists[178] because they "produced counter-environments as a way of critiquing them" (McLuhan and Cavell 84).

The methodology of using counter-environments rendered by artists started at the beginning of McLuhan's career. In order to bridge the gap between himself and his students, he selected sources from popular culture, mainly advertising, to provide a window into industrial culture and, by comparison, to illustrate a particular understanding of human experience. It was a praxis that had been gaining in popularity during his time at Cambridge and practiced by poets such as James Joyce, T.S. Eliot, and Ezra Pound (McLuhan and Stearn 50–51). By taking these items out of their familiar context, for instance by juxtaposing art or literature with an advertisement or a comic strip, he wanted to jolt his readers and listeners into a reexamination of what would otherwise be familiar and thus often imperceptible surroundings.

No Classification without Misrepresentation

Clearly, McLuhan's aesthetic approach to understanding technology does not match up neatly with objective and reasoned scientific modes of presentation. Coupled with the professorial flair in his prose and the dogmatic tone of his statements, it makes McLuhan's ideas unmistakably prone to misconception. One only need to observe his speaking and writing style, which tends to be figurative and nonliteral, more exploratory than explanatory (McLuhan and Gordon 10).[179] In

---

"Personally, having found the utmost ambiguity in all human technologies, and having never discovered a fixed position from which to view or measure them, I have settled for studying their on-going effects on the users." (Gordon, *McLuhan: A Guide for the Perplexed* 135).

177 In a chapter on television in *Understanding Media*, McLuhan does not seek a solution through an understanding of the power of technology. Mediated reality, or "the pattern of experience presented" helps shape culture and entices conformity to societal norms (McLuhan and Gordon 436).

178 As James Curtis points out, McLuhan "used art, not documents, as primary evidence in historical analysis" (149).

179 In a letter to William Kuhns, McLuhan writes that he satirizes "at all times" and that his "hyperboles are as nothing compared to the events to which they refer" (*Letters of Marshall Mc-*

contrast to gaining knowledge through scientific ordering and classification, McLuhan preferred the probe as a method of inquiry. He once explained his approach as follows:

> Classification, for the literary man, is the be-all and end-all of observations. . . . In the medical world, classification is a form of dismissal. If the doctor says it's measles, that's it, it's over with. The rest is just routine. But classification is not the beginning of the study of a problem–it's the end. For me any of these little gestures I make are all tentative probes. That's why I feel free to make them sound as outrageous or extreme as possible. Until you make it extreme, the probe is not very efficient. Probes, to be effective, must have this edge, strength, pressure. Of course they sound very dogmatic. That doesn't mean you are committed to them. You may toss them away. . . . I'm perfectly prepared to scrap any statement I ever made about any subject once I find that it isn't getting me into the problem. I have no devotion to any of my probes as if they were sacred opinions. I have no proprietary interest in my ideas and no pride of authorship as such. You have to push any idea to an extreme, you have to probe. Exaggeration, in the sense of hyperbole, is a major artistic device in all modes of art. No painter, no musician ever did anything without extreme exaggeration of a form or a mode until he had exaggerated those qualities that interested him (McLuhan and Stearn 54–55).

Central to such an aesthetic approach are symbols and metaphors, not definitions or classifications. Similarly, accuracy and objectivity are not of primary interest to a scholar using this methodology. McLuhan may have understood that meaning is best conveyed through such literary devices; that they can be used to unify and make sense of experience; that one can gain more understanding of technology from a metaphor (technology = media) than by defining or classifying it ad absurdum. As his academic career progressed, McLuhan used aphorisms or short phrases like the popular "medium is the message" and the "global village"[180] to express a unique and complex idea and to provoke reflection and fascinate his reading and viewing audiences. McLuhan's way of probing and observing the fallout can be viewed as akin to trial-and-error methodology. If an audience is not prepared for or trained in knowledge of literary devices, his poetic license to formulate and then kill off an idea that does not work can obfuscate a deeper interpretation of his work.

The fact that he often chose to emplot the ideas in his writings within a framework of necessity further adds to reader confusion. In his most widely read

---

Luhan 448). McLuhan emphasizes the conscious use of his writing style by declaring that "I can write the ordinary kind of rationalistic prose any time I choose to do so." This technique of symbolism McLuhan calls "rip-off" and used it to "deliberately deprive" figures of their ground.

180 The global village metaphor is easily misconstrued to imply mutual understanding. McLuhan, however, used the metaphor to indicate that, for better or for worse, through electronic communication, everyone could participate in a conversation.

work, *Understanding Media* (1964), McLuhan used discontinuous prose, and as a historian-author of the technological past, he narrated a past using an epochal model of media that conveys elements of necessity. Even a careful, close reading can be deceiving, leading to a singular and linear comprehension of the cause-effect relationship between technology and culture in which, for example, the printed word (in different languages) inevitably created the one effect of nationalism (*Understanding Media*, "The Printed Word: Architect of Nationalism" 231). Such a narrowing of possible outcomes appears to undermine an understanding of the overall framework of contingency between technology and society. And as James Curtis has noted, such simple reasoning is deeply problematic even for traditional historians:

> Historians often object to such unilinear reasoning, asserting that such complex phenomena as the Renaissance always have multiple causes. I suspect that this type of objection stems from unconscious reliance on a geometrical model of vector analysis. Just as any given vector represents the resolution of a theoretically infinite number of forces, so any given event represents the result of a large number of causes in historical analysis. We can readily visualize this image, just as we can visualize the lengthened shadow in Emerson's definition of an institution, but it has no logical force in and of itself. Many criticisms of McLuhan lose much of their cogency when we recognize that he did not use visual models for historical processes (148).

After the publication of *Understanding Media*, McLuhan's status changed from obscure to famous. Once the book's readership had gone beyond its disciplinary boundaries, its author's aesthetic approach to the subject invited a multiplicity of interpretations and critiques from prominent scholars.[181] Perhaps the irony of a rise in popularity is being misinterpreted, a development which may have contributed to his publisher requesting a reformulation of his ideas in more "scientific" terms. The resulting book eventually received the title *Laws of Media*, was co-authored with his son, Eric McLuhan, and published posthumously in 1988.[182] Some twenty years later, Eric published another attempt to summarize his father's theoretical stance and to answer questions about his supposed theory of communication. In chapter 16 of *Theories of Communication* (2011), the younger McLuhan provides a brief response using a directly quoted statement from his father.[183] The short answer is no, he did not have a theory of communication. In the more elaborate ver-

181 For a summary of the critical reception to *Understanding Media* see McLuhan and Gordon 454–558.

182 A collaborative work between father and son, *Laws of Media* presents the tetrad as a means of exploring causality through the mapping of media effects.

183 Eric McLuhan starts the chapter concisely by answering the question of whether Marshall McLuhan had a theory of communication: "Whenever provoked, Marshall McLuhan would de-

sion of the answer that follows, he provides numerous examples of his father's methodology, describing his probes, observation, aphoristic language, training perception, etc. (171–87).

In sum, McLuhan's ideas often defy neat description. He denied having a theory to guide his research, employed aesthetic methodology and scoured mainly literary (as opposed to historical) sources for his writings, and especially towards the later stages of his career, used markedly inflated prose. He avoided working directly with a broad term like technology, and the implications of his research into the relationships among technology, history, and society provide a different perspective as opposed to a representation (medial time or various periods of time in which there is a dominant media environment for human communication). He might point out that this is how the probing methodology is intended to work. But to gain insight into McLuhan's contribution to discourses on technology in the twentieth century, an understanding of his methodology and sources is only a starting point.

Through his readings of Mumford and Giedion, McLuhan was aware of frameworks to write about a technological past that was understood to be contingent and anonymous. This is one reason why, instead of insisting on the necessity of historical development, his account of the past traced changes backward, starting with effects. While he did narrate technological change through major shifts in human communication, he made no attempt to claim any type of predictability in history beyond recognizing that causes were often preceded by effects. In order to situate his writings in the context of a shift in scholarly discourse and historiographic practices, the current analysis finds evidence in two relations to the past – aesthetic and epistemic. Distinguishing these two specific relations in McLuhan's work helps to explain his guiding aim in constructing a more coherent account of the past. By offering a grand narrative of medial changes, McLuhan argues for reference to what Aristotle called "formal causality" as a theory of technological change and advocates the importance of understanding the effects of past communication technologies. As a service to humankind, the awareness of a collective medial past can be used to question the current, dominant media environment and is the key to countering its effects. The following passages attempt to contextualize these relations using a selection of his writings, both early and late, with examples from the texts.

clare, look, I don't have a theory of communication. I don't use theories. I just watch what people do [. . .]" (*Theories of Communication* 171).

Combatting Dissonance with Coherence – An Aesthetic Relation to the Past

A major shift for McLuhan personally occurred when he moved from the U.S. to Canada in 1944. After two years at Assumption University in Windsor, in 1946 McLuhan became one of the first laypersons to join the teaching faculty at St. Michael's College in Toronto.[184] In 1952, the same year he became a full professor, he published a short article which represents an early attempt to "superficially" examine the topic that would dominate the next decades of his academic career. In "Technology and Political Change," McLuhan argued for an increased study of how uses of technology impact culture and politics. His analysis throughout the text examines the effects of past changes in communication, which ultimately led to a specific political expediency for current times – an awareness of a positive "by-product of new techniques of communication," namely cultural equality. In his view, it allowed for exchange and autonomy amongst societies. Once established as a value, as McLuhan put it, cultures are then "free to learn from all the others while possessing [themselves] in quiet" (195). The question is, how does McLuhan link past changes in uses of technology to changes in politics and arrive at such a conclusion?

In the opening claim of the article, McLuhan reminds his readers that all technologies of communication have their own aesthetic modes and that the mode itself is a vital part of the act. This forerunner to the later well-known paradox "medium is the message" preconfigures his narrative in two important ways. First, it assumes an aesthetic relation to the past in which the "story" is more concerned with form than with content. And second, it allows McLuhan to colligate the past under one theme and thus achieve a degree of narrative closure.[185] Once communication and its form have been established as a coherent way to frame the past, McLuhan is able to break up time into medial epochs. He does this often using a single word like writing, which can be understood as a literary device (synecdoche) to represent the totality of the social and material processes when a particular medium was the dominant form of communication. One early passage in particular is indicative of how his medial/epochal framework aims to foster an aesthetic understanding of the past. If read literally, however, it seems reductive and insinuates the notion of technology as a driving force, which can be understood to involve linear causality and a framework of necessity. As a form of communication, McLuhan mused, writ-

184 See entry "Our History" in the works cited section at the back of the book.

185 An aesthetic relation to the past implies a certain narrative template ("story form") that helps a historian to frame historical thought. According to historical theorist Herman Paul, narrative closure is when historians are able to achieve "coherence of meaning" in a story (Paul and Runia 56–57).

ing reduces the quality of an otherwise spoken message and extends its range across time and space. Whereas the spoken word is used for effect on an audience, the written word is meant for private consideration. Judging by the extended range of possible communication and the decreased amount of people able to read the message, McLuhan drew the conclusion that writing was a political revolution. In other words, the effects of the changes in form of communication were changes to social communication and control (189).

Delving further into his analysis, McLuhan noted the heightened importance of the subsequent mechanizations of both writing and speech. As an author-historian, McLuhan applied his aesthetic relation to the past by describing mechanization as a change in form that has traceable effects. Referring to this change in form as printing, McLuhan claimed that "the mechanization of writing reduced the effect of the spoken word even more than had the invention of writing" (190). His evidence includes effects like the rapid dissemination of books, which he believed aided the transition toward more private modes of teaching and learning. He stated in rather stark terms that, "Politically speaking, this social change was felt in the new intensity of commercial exploitation of vernaculars. Printing fostered nationalism when the printers sought to extend their markets as widely as possible." It is important to observe that his use of single words (writing/printing), as is evident in the sentence quoted above when he includes the printers themselves, always implies the co-development of technology and society – both human ways of perceiving and acting using a particular medium of communication as well as the effect of the form on the content and content-creators themselves. McLuhan also hinted at the contingency that framed his understanding of technological change, stating that the increase in speed and decrease in cost of the mechanization of writing led to "a whole series of unpredictable social and political consequences" (190).

Having spent time emphasizing the effects of changes in communication in the distant past, McLuhan switched to probe more contemporary forms: the book and the newspaper. Aesthetically, he argued, books "abridge" time. This means that a book, as a mode of communication, connects an author to a reader when its contents are received. In contrast, newspapers, or "new book of the people" as he called them, collapse space and render time before the date line insignificant. In other words, he argued that, aesthetically, an old newspaper is not interesting and knowledge production tends to revolve around the present. McLuhan added that newspapers enable stories about geographically disparate locations to coexist on the same page and help create "an impact of immediacy and super-realism" for that one day in particular (191). With a hint of sarcasm McLuhan admitted that "naturally, the technologically determined format" of newspapers, along with its enablers in the press, "had revolutionary political consequences. It has changed everybody's way of thinking, seeing, feeling" (190).

One of the ways that McLuhan argued that experience of the world was changing in the late seventeenth and early eighteenth centuries was in seeing oneself as part of the scene and, taken to its logical extent, as part of history. A metaphor in everyday language was given as evidence. McLuhan pointed out that using views "as a way of expressing moral and political attitudes" is closely associated with the rise in newspapers (191). Also encouraging this "spectator mentality" were concurrent inventions like the telescope, microscope and spectroscope and the *camera obscura*. McLuhan's preferred account of the past thus associated changes in communication form with innovations and, ultimately, with effects in culture. But buried in this at times vague amalgamation of explanations for technological change is the insight that cultures can be viewed as a communication network. This means that forms can precede causes that eventually affect other parts of the network. The example he provided is the form of the detective story. He argued that it preceded modern science/historiography "in the discovery of this method" in which an effect can be traced backwards and reconstructed to fit a story perfectly (192).

McLuhan lamented what he saw as the many overlooked consequences of the mechanization of writing, "the real power of these deities is exerted when we aren't looking" (193). He called for more attention to be paid to the effects of "revolutions" in communication. Increased scrutiny of dominant forms of media is especially important at a point in his aesthetic framing of history, when the extraordinary is made trivial through the mechanization of speech. Such a change in form of communication has had the effect that ". . . all the networks of human communication are becoming so jammed that very few messages are reaching their destination" (193–194). Ultimately, McLuhan wanted his readers to understand forms of communication amidst a cacophony of voices to avoid the "mental starvation in the midst of plenty." Such emancipatory awareness of form, the argument goes, might help them from being squeezed into certain modes of thinking or desires through modern advertising. To indicate a possible solution, he cited not a historical source, but a literary one: ". . . if the new reality of our time is in the main a collective dream or nightmare brought about by the mechanization of speech . . . then we must learn the art of using all of our wits in a dream world, as did James Joyce in *Finnegans Wake*."

McLuhan saw the latest development in communication as yet another revolution. Now that modern devices allow for instant communication, the result is a social, historical, local, and temporal simultaneity. He claimed that electronic forms of communication had fundamentally changed communication once again, or as Harold Innis might put it, the things to which we attend. Specifically, it has significance for the study of the past because what he later called a "revolutionary state of mind" had managed "to lessen the romantic appeal of distant times and customs in favour of a direct stylistic interest in their immediate value and relevance" (194). He continued this line of analysis: "the modern study of the past, as of distant places, has the effect of making them as much a part of the present as our own problems. So that for the modern mind, history has not become a receding perspective but a present burden" (194). To combat the disorienting and distressing effect of modern communication technologies McLuhan did not point to television or radio or any other material technology, but rather to monumental ideas with great political repercussions.[186] He noted that "the great political discovery of the eighteenth century was social equality. The principle insight of this century to date is perhaps the anthropologist's awareness of cultural equality." The latter insight, cultural equality, was an effect of changes in communication.

Keeping in mind the platform of the article, the *International Journal*, McLuhan's view on international affairs appears to recommend that solutions can be found amongst the newest changes to communication. Once one recognizes that various cultures can be understood as communication networks equal in their ability to respond to "specific problems in interpersonal and inter-social communication," the subtext to McLuhan's article becomes more lucid: a ground for political relationships on equal footing. McLuhan was well aware of the effects of American news and advertising on the Canadian mind. An implication of his analysis is that Canadian culture was in just as viable a position to shape its own as well as global political and cultural policy as the colonial power to the south. In other words, "the impact of technology on culture and politics" should not only be understood as [American] technology driving history, but that changes in form of communication also allow for the emergence of "benign" forces like the awareness of cultural equality, which can have an effect on "practical politics" (195). This reading renders the conjunction in the title of the article all important, technology *and* political change, two forces bound in a network of mutually enabling/disabling relationships.

Synthesizing the Medial Past – An Epistemic Relation to the Past

A few years after "Technology and Political Change," McLuhan was asked to contribute his ideas about the relationship between technology and society to a collection on the craft of printing. Instead of taking on the role of an informed

186 Also appearing at the beginning of the essay, no matter how much technological change helps connect people to others and to knowledge, it does not affect their "metaphysical isolation" ("Technology and Political Change" 189).

Canadian for a political journal,[187] a more historical platform allowed McLuhan to assume the role of historian-author. Entitled "Printing and Social Change," the article appeared in *Printing Progress: A Mid-Century Report* and represents a deep dive into one particular mode of communication. It traced the social effects of print, or the changes brought about by print culture by "consulting" the past 500 years of what he referred to as the "Gutenberg Era." Although the article can be seen as a continuation and a crystalizing of the periodization of medial forms previously undertaken, McLuhan now wrote unapologetically from the perspective of the "Electronic Age" (89–90). A stauncher approach to breaking with the past was an outgrowth of his years as an editor of *Explorations* (1954–1959). This collaborative journal on studies in culture and communication, funded by the Ford Foundation, represented a shared space for McLuhan to develop his ideas on new media culture with scholars from other fields.

In "Printing and Social Change" McLuhan explained the profound effects of print in an attempt to advance a pedagogical vision for what he viewed as a new age of media. At the end of the article, McLuhan argued that learning should be done by exploration as opposed to by classification to "meet the demands of a new age" (112). Analyzing his epistemic relation to the past, meaning a relation that is centered on "knowledge and understanding of the past," McLuhan can be recognized to have used his account of the past to cultivate an understanding of the present (Paul and Runia 84). To do this he made a number of generalizing inferences about the last five hundred years of communication media. For example, at the beginning of the article, one such inference is his characterization that a feature of the printed word can be defined as a "still shot" or "an arrested moment of mental movement" that fosters a way of thinking ("Printing and Social Change" 89–90). McLuhan wanted to question the qualitative effects of print. He was not concerned with how the mechanization of writing into print affected the number of books produced (quantitative), but rather how the "vitality of new forms" in the Gutenberg Era affected the existing patterns of knowledge and perception (90). How did these forms reshape human sensibilities? McLuhan described print as a "complex new situation," which resulted in "an analytic habit of awareness" that encouraged a "static individualistic perspective" (96–97). He eventually provided some names for the effects of print, calling one "lineality," and attributing adjectives like "analytic, visual and consecutive" to print (103, 110). This way of thinking, he wrote, had led to a common understanding of change during

187 The *International Journal* was started "to provide a platform and outlet for informed Canadian views on international affairs" (See entry "Our History" in the works cited section at the back of the book).

the Gutenberg Era that he called efficient causality. It is akin to printed words on a page in which one item follows another in sequence: A causes B. And McLuhan added that it had become a concern of the last few centuries.

Throughout the text, McLuhan used a variety of words and phrases to interpret technological change. A close reading of the text identifies McLuhan's description of historical change as a dynamic process in which ". . . any pattern of knowledge or of perception is affected by new information and by other co-existing patterns" (90). At the very least, McLuhan's own understanding of technological (medial) change included not only sequences and advances, but also "unexpected reversals of form," of which "any history of technology is filled" as well as translations of experience from one form to another (91).[188] He valued the concept of efficient causality for its contribution to science and learning, but made clear that this understanding had been superseded by the new media environment.

Deeper into the text, McLuhan made his synthesizing inference about the medial past more explicit. He suggested that the printed word and the movie form the parameters of a "process of development" in the "codification of reality" or "new way of representing and communicating mental activity" (95). But this claim is qualified when he stated that the complex process behind such changes cannot be attributed to a "solitary invention," but rather to a "cluster of many technological achievements which had accumulated over the centuries." It becomes apparent that McLuhan was challenging the notion that technological change is unilinear and predictable. Despite acknowledging that society had both benefitted and suffered from a long period of literacy that shaped consciousness, McLuhan argued that the rise of electronic media (telegraph, telephone, radio, TV) now presented a unique and novel sensory environment and required "research into the new situation" (111). He was interested in the new effects in individuals, which, if his account of the medial past is to be taken seriously, would ultimately affect perception in a fundamentally differently manner than print media.

What McLuhan did in these "very brief notes on a very complex subject" was to re-think the past in medial terms (107). He identified effects and habits of mind that had supposedly arose in the so-called Gutenberg Era, a colligatory concept he used to create order in his story about the past.[189] In the process, he also detailed a specific understanding of causality that he associated with print – efficient cau-

188 One example of how McLuhan sees media as translators of experience is the phonetic alphabet. As a technology, it translates the audible into the visual.

189 British Philosopher William H. Walsh used the term "colligatory concepts" to refer to words that historians use to group people, places and happenings into one concept to help them "create order in their stories about the past" (Paul and Runia 120). Other examples are the 'Cold War,' the 'Renaissance' and in McLuhan's specific case, the 'Gutenberg Galaxy' and 'Electronic Age.'

sality. McLuhan thus provides his readers with the knowledge that the past can be narrated through the effects of media. But more pertinent to the overarching trend towards a more contingent framework to understand technological change, he used this story about the past to give a plausible explanation as to why efficient causality was no longer paramount. It is only one possible understanding of technological change.

Theory – A (Book of) Concern with Past Changes to the Patterns of Human Perception

McLuhan soon expanded on ideas about the effects of literacy and print media as detailed in "Technology and Social Change." In writing about the origins of what he called the "Gutenberg configuration of events," he sought to provide his readers with more in-depth scholarly evidence on the "older forms of perception and judgement" that had been built up over the past 2000 years (*The Gutenberg Galaxy* 18; 278). *The Gutenberg Galaxy* (1962) was written in mosaic form, which was central to McLuhan's philosophy of revealing causality in a historical account. As briefly alluded to on the page opposite the contents section, a mosaic approach meant that he considered the data and quotes used in the book to be part of a participational or interactive analysis of the past. It was an Innis-inspired employment of configurations as opposed to an explanation of event sequences (216). In this contingent, networked understanding of historical causality, forms undergo transformation and lead to different constellations. McLuhan later described the book more modestly as "a series of historical observations of the new cultural completions" that had resulted from the "disturbances" of literacy and then printing (4). As a theoretical grounding, McLuhan reformulated an observation by Romantic poet William Blake that new technological forms result in new perceptions: "sense ratios change when any one sense or bodily or mental function is externalized in technological form" and "when sense ratios change, men change" (265). Essential to understanding the changing patterns of human perception, according to his argument, is awareness of the "inevitable [human] drive for 'closure', 'completion' or equilibrium" that is made through extension or suppression of the senses in response stimulants in the environment (4).[190]

With regard to tensions between modern and postmodern historiography, employment of an astrological trope to describe the book's subject is significant.

190 Elsewhere McLuhan describes this closure as a "novel pattern of interplay" among the senses (23).

Figurative use of the word galaxy demonstrates how McLuhan chose to symbolically mediate a cultural interpretation of the past to his readers. In doing so, McLuhan argued for a shift from the search for linear causes in historiography to an analysis of a certain galaxy or configuration constructed through interpretation by the author (252). And because he thought that "a theory of cultural change is impossible without knowledge of the changing sense ratios effected by various externalizations of our senses," the metaphor of the galaxy allowed McLuhan the flexibility to associate a variety of textual nodes in his account without needing to correspond to the truth (42). Emphasizing that his role as an author-historian was to observe and understand change, but not to judge it, he criticized taking on a "point of view" for his study because he thought that, methodologically, doing so would amount to a "closed system" that he affiliated with the "unconscious effects of literacy" (6, 213).

Delving into the specifics of the effects of changes to communication, McLuhan began the book by claiming that interiorization of the technology of the phonetic alphabet had translated sensibilities of the Greeks from "the magical world of the ear to the neutral visual world" (18).[191] Thereafter he broadly followed "the ways in which the forms of experience and of mental outlook and expression have been modified" by manuscript, and then by print culture (1, 25). This exercise in textual history seemingly culminates when McLuhan expressed, in figurative language, what can be easily misinterpreted claims about the causal relationship between nationalism and print culture:

(1) "Print, in turning the vernaculars into mass media, or closed systems, created the uniform, centralizing forces of modern nationalism" (199).
(2) "Closely interrelated, then by the operation and effects of typography are the outering or uttering of private inner experience and the massing of collective national awareness, as the vernacular is rendered visible, central, and unified by the new technology" (199).

Despite the deterministic overtone in these statements, the central claim of the book is broadly that print culture encouraged a more isolated, visually oriented experience of the world, or as McLuhan put it, "typographic logic" created "typographic man" (212). In crafting these ideal fictional typologies, in no way did McLuhan cement himself to historical timelines and causal descriptions or preclude other factual interpretations of the past. Instead, his tracing and narration of a historical transition from

191 Looking backwards to find the origins of the effects of literacy, McLuhan technically defines a man as "civilized" when "visual values have priority in the organization of thought and action" (*The Gutenberg Galaxy* 27).

speech to writing and then print was based on an understanding of historical time or human experience that is non-linear.[192] Historical media culture in the above sense can therefore be understood as a constellation which develops from various reactions to new technologies. On this view, communication takes on a dominant form that can reconfigure perceptions and affect attitudes, beliefs, and values. The more significant change, McLuhan maintained, is not necessarily during the introduction and dissemination of a new technology, but rather throughout a long period of adjustment of "personal and social life to the new model of perception set up by the new technology" (23). This more multifaceted and contingent framework for social change justified McLuhan to assert that, if understood correctly, "far from being deterministic" his work should "elucidate a principal factor in social change which may lead to a general increase in human autonomy" (3). That principal factor is of course formal causality, a concept that he contrasted with the efficient and sequential causes perceived by someone who has been immersed in the "abstract explicit visual technology of uniform time and uniform continuous space" (19).[193]

In an attempt to teach an awareness of forms and formal causality, McLuhan's account of the medial past constructs historical knowledge of the "translation or reduction of diverse modes into a single mode of homogenized things" to actually help "restrict areas of determinism" (261, 247). He claimed that "emancipation from that trap [determinism] is the goal of all education." Conversely, he warned that being uncritical of the influences on patterns of thought derived from mechanical technology would lead to "maximal determinism in human life" (247). McLuhan called for an escape from mechanical culture and ways of thinking. At the end of the book, he thus pivoted from a rendering of the past (*Gutenberg Galaxy*) to set up a study of the present (*Understanding Media*).

Praxis – Emancipation from Mechanical Modes of Perception

Before and during the publication of *The Gutenberg Galaxy*, McLuhan was at work extending his theoretical conjectures about the mechanical past into the organic present though experiment and praxis. The book that eventually became *Understanding Media* had its foundation in a research project supported by the U.S.-based National Association of Educational Broadcasters (NAEB) and in the subsequently filed *Report on Project in Understanding New Media*. McLuhan

192 Analyzing a then recent book on Italian theorist Giambattista Vico's theory of history, McLuhan writes that Vico understood "the time-structure of history as non-linear, but contrapuntal" (*The Gutenberg Galaxy* 250).
193 McLuhan's ideas about formal causality are also elaborated in *Media and Formal Cause* (2011).

summed up the goal of the report by stating that he had thought it necessary "to scrutinize modes of media, in order to reveal the assumptions they impose and the perceptual habits they foster" (McLuhan and Gordon 479). The *Report* also included a sample syllabus for teachers. Here he further affirmed the above goal, professing that by studying the "action of media" on perception and judgement, one could "escape into understanding" and "success" and remain "reasonable and autonomous" (516). According to an early reviewer of the syllabus, McLuhan would need to better introduce and explain the word "media," as it was not a common term at the time (478). McLuhan responded to that criticism by expanding the number of media that he had studied in the *Report*. Moving from a report on research findings to an interpretive, educational guide to media, McLuhan transformed the *Report* into a monograph over four years "by raiding his own library" (539). But, as W. Terrence Gordon noted in the works cited section of the critical edition of *Understanding Media*, the writings of I.A. Richards and Harold Innis are notably absent from the references of the book (569). This is because, as established previously, the ideas of New Criticism (Richards) and Staple Theory (Innis) fundamentally informed the underpinnings of McLuhan's entire theory of media.

McLuhan chose a title and framework for *Understanding Media* that he had modeled on academic textbooks *Understanding Poetry* (1938) and *Understanding Fiction* (1943).[194] These two latter books informed McLuhan's structuring of *Understanding Media*, which he broke into two parts. A quote from a later edition of *Understanding Poetry* clearly illustrates what McLuhan was attempting to describe in Part I of *Understanding Media*. The quote from *Understanding Poetry* is as follows: "the title of this book is *Understanding Poetry*. It might, however, with equal reason have been called *Experiencing Poetry*" (15). Similarly, the first part of *Understanding Media* describes what it is like to experience media and what should be done to counteract their effects. Part I also forms the core of McLuhan's theory of media. With a few exceptions, the first chapters reiterate themes that had been expressed in other writings, such as media being the extensions of the human senses, an overextension of a medium causing a reversal, that individuals seek out equilibrium, the need for antidotes for the effects of media, and the concept of media as translators.[195] While explaining the effects of media on the senses, McLuhan evoked the myth of Narcissus and indicated the common misunderstanding that Narcissus had fallen in love with himself. Conversely, it was his "failure to recognize his image that was his undoing. He succumbed to the typical numbing effect

194 Both books can be considered part of the New Criticism movement in literary theory.

195 For example, from *The Gutenberg Galaxy* that print translates "the dialogue of shared discourse into packaged information" (164).

of any technology, as it lulls the user into a trance. Technologies create new environments, the new environments create pain, and the body's nervous system shuts down to block the pain" (McLuhan and Gordon xvii). This story thus encapsulates McLuhan's argument, that media, as extensions of our bodies (which also include mind) are creators of environments that affect and disorient our sensibilities. This is especially the case with the advent of electronic media, where the nervous system is constantly bombarded from all directions, not merely by a dominant media form, but by different media forms.

A quote from *Understanding Fiction* suggests what McLuhan was attempting to accomplish in Part II of *Understanding Media* – not to define technology as a concept, but rather to analyze singular instances of media. Avoiding a definition that is either too simple and unreliable or too complex and abstract, the authors of *Understanding Fiction* had preferred to "work toward an understanding of our subject through an investigation of particular instances" (1). Accordingly, Part II of *Understanding Media* discusses the application of the thoughts expressed in Part I in relation to individual media. There are sections on the spoken, written and printed word, on TV, on the telephone, but also on non-traditional media like chapter 19 on the wheel, bicycle, and airplane.

On the Relationship of Human and Media Environment – A Material Relation

In addition to an aesthetic relation and an epistemic relation to the past, analysis contained in *Understanding Media* investigates a material relation to media environments. Serious consideration of this relation has opened up several interesting avenues of research,[196] but also has had particularly unfortunate consequences when interpreting McLuhan's understanding of technological change. In the book, McLuhan probed the physical effects of technologies on the human senses and sensibilities (McLuhan and Gordon xi). This acknowledgement of the materiality of technology as an analytical starting point has led to McLuhan's work being miscast as technologically deterministic. On a different interpretation, by first concerning himself with physical effects, McLuhan can be characterized as merely having emphasized a material relation to a media environment. As historical theorist Herman Paul reminds his readers:

196 One of which is Armond Towns's research on "McLuhan and the Racial" in which he posits the black body "as one medium through which concepts of man or the human are extended" (talk at the "Many McLuhans Symposium" on 21 September 2018, https://www.mcluhancentre.ca/manymcluhans: accessed on 05.06.2019).

> unlike other relations [with the past] . . . the material one is not characterized by teleological activity. . . . it is the relation of indebtedness and dependency. It revolves around what the past does to people, rather than what people do with the past. The material relation refers to how people are influenced by the past – where and in what circumstances they are born . . . . In a sense, therefore, the material relation precedes all other ones (Paul and Runia 34).

McLuhan's particular understanding of the relationship between technologies and environment hinges on his Innisonian, meaning staple interpretation of media – past, present, and new. He refined his analysis of media to understand them both as extensions of the human body and "forms of social organization and interaction" (McLuhan and Gordon 34–35; 564). Theorized in this way, they can be interpreted both as "natural resources" that have "formative power" to configure "the awareness and experience" of individuals, but also to effect "social patterns of organization." Such broad material causality in his thesis is responsible for some scholarly critiques of his work. Understood within a mechanical framework, media's material relation seemingly imbues it with strong historical agency, making it a driving force and thus increasing the relative degree of necessity in causality. Taken to its extreme form, as McLuhan was apt to do, led to formulations of pithy statements like print caused nationalism. What is implied is that print culture, which had stabilized over centuries of interaction among various social and material forces, became a fundamental environmental factor that had channeled ways of thinking and interacting. In this scenario, an individual, like a ball inside a bowl, requires great effort to move up and out of its container.

Ironically, part of the effort to break out or change from a dominant media environment and its effects cannot occur at the conceptual level. That is because, as McLuhan stated, media "alter sense ratios or patterns of perception steadily and without any resistance" (31). While the various media represent the human body's effort to increase its capabilities, these extensions do not bring about a corresponding increase in perception. In fact, McLuhan argued that the effects of certain technologies actually hinder the ability to perceive one's environment. He elaborated on the equally important, but less attractive "other" to extensions, so-called "amputations," which are the result of a numbing effect produced by a dominant medium. Such isolation results in a lack of self-recognition, of perceiving the ways in which human extensions shape the world. McLuhan described this effect as suffering from "somnambulism" (23). One is less aware of a technology and its effects and the psychic and social adjustments that need to be made to adapt to it. It is here that McLuhan invoked his famous parole, "the medium is the message,"[197] to explain the "change in scale, pace or pattern" that is introduced to "human action and interaction" by new forms of technology (564).

A Study of History – New Media Environment as a Challenge to Culture

McLuhan advanced his concern with a cultural understanding of technology and its current effects by amplifying their historical significance. He inferred that, seemingly unaware of the cultural consequences of technologies on individuals in society, the response by conventional historians had been to become numb to them. By way of example, in *Understanding Media*, McLuhan criticized the great British historian Arnold Toynbee for a lack of understanding of "media as they have shaped history" (31). Despite the deterministic prose, McLuhan was attempting to convey the historical neglect of the type of technological change that he had sought to understand. The fifth chapter of the book is rather instructive in this respect. Describing conventional understanding of technological change, McLuhan noted that "everybody notices how coal and steel and cars affect the arrangements of daily existence," yet few were aware of the "medium of language itself" as a major factor in shaping social realities (74). This is because media like language, are creators of new environments.[198] They are agents that "make happen," but do not "make aware." And they are not alone in shaping media. In the same paragraph, he was quick to include the role individuals play in the process, "being extensions of ourselves, also depend upon us for their interplay and their evolution" (73). These passages clearly demonstrate that McLuhan was an advocate of the co-creation of media, yet his analysis went beyond their genesis and addressed the issue of further control. McLuhan answered that once one was aware of media's effects, one could control them: "we can, if we choose, think things out before we put them out" (72–73). In McLuhan's mind, it was foremost the perception of medial environments that is the key to understanding media, control of their effects, and thus greater human autonomy (76).

In *A Study of History* (1948), Toynbee detailed how a number of cultures had responded to change, a sort of challenge-and-response framing of history. In his own historical analysis, McLuhan embraced this framework to focus on the chal-

---

197 Originally printed as "The Medium is the Massage." See McLuhan, Eric. "Commonly Asked Questions about McLuhan" in the works cited section at the back of this book.

198 With regard to technological change, it is important to point out that McLuhan considered language to be humankind's first technology. Understood in this sense, language makes it possible to let go of an environment in order to grasp it in new ways.

lenge posed to all cultures by electronic media environments. If one were to embrace the cultural response frequently endorsed by the British historian, it would call for a reliance on special individuals and resolve. But McLuhan thought this strategy depended too much on copying the actions of great personalities and seeking comfort in the power of the will to appropriately respond to cultural challenges. McLuhan's own response strategy combined two great "discoveries" of the past two centuries to help him in understanding media, which again, for him was inseparable from culture: the "technique of discovery" itself and the "technique of suspended judgement" (*Understanding Media* 93). The former he described as "starting with the object to be discovered and working back to the point at which it is necessary to start in order to reach the desired object," while the latter is unique in that it can anticipate an effect and employ countermeasures before it happens. In other words, McLuhan sought to work around the need to adjust (unconsciously) or create adequate responses on-the-fly through an understanding of environments and an anticipation of their effects.

To put this strategy into practice, McLuhan insisted on the "power of adequate *perception* of situations" (102). He lamented a general lack of a "conscious adjustment of the various factors of personal and social life to the new extensions," and affirmed his conviction that artists, in the broadest sense of the term, were the ones to lead the way. McLuhan claimed that artists' "shaping and analysis and understanding . . . of forms, and structures created by electric technology" was vital to successfully adjusting to technological change (95–96). In addition to an ability to grasp the current material, medial environment, McLuhan added that artists could also rebalance sense ratios before the numbing psychic and social effects of new technologies would occur. The theory section of *Understanding Media* culminates when McLuhan issued a warning to reinforce the importance of human awareness in responding to changes in the technological environment: "As long as we adopt the Narcissus attitude of regarding the extensions of our own bodies as really *out there* and really independent of us, we will meet all technical challenges with the same sort of banana-skin pirouette and collapse" (100). Without control over new media, Toynbee's challenge-and-response narrative was reformulated in *Understanding Media* as *Challenge and Collapse*.

The Epistemology of Causality – A Formal and Dynamic Concept to Explain the Relation Between Cause and Effect

Admittedly, one may be forgiven for being confounded by an explanation of causality that appeals to an abstraction like the so-called extension thesis, which maintains that technology is the extension of the human body (Lister 89). But the essential point that McLuhan was attempting to argue was that causality, beyond being conceived as linear and sequential, can also be understood in formal and dynamic terms. Along with colleague Barrington Nevitt, McLuhan provided a characteristically dense, but rich analysis of a less conventional conceptualization of causality in a 1973 article entitled "The Argument – Causality in the Electric World," which was printed in SHOT's journal *Technology and Culture*.[199] Referring to Harold Innis's observation that there had been a shift in dominant media and modes of perception, which involve a sensory bias, as the Greeks became less audile-tactile and more visually-oriented, the authors argued that Western Culture with regard to literacy had created a dominant understanding of causality. Specifically, they suggested that through an increasing use of writing and the alphabet, Western civilization had been encouraged to separate subject from object and thought from feeling. This situation was compounded by what McLuhan had argued elsewhere, namely that the dominant metaphors of mechanism had widened the perceived gap between cause and effect (3).

A new "causal reality" had emerged over time due to a shift in the environment, so-called *ground*, in which changes could be perceived. Citing a number of scholars that include Karl Popper and Thomas Kuhn, McLuhan and Nevitt exposed the limits to using such a mechanical (linear, sequential) concept of causality to explain change in the electric world. This included critiquing both its use to explain the past and predict the future, but even more so as a "program" or agenda for scholars to pursue. In its place, the authors advocated using a formal, dynamic concept of causality "as a probe for new knowledge" that "re-cognizes" or rethinks the manner in which knowledge comes into existence in an electric world (5–6). To illustrate the difference between the mechanical and dynamic understanding of the concept, they stated that in the nineteenth century, "reality" was akin to matching the old. In a historian's case, this meant finding and justifying causes and connections for what had already happened, or matching theorized causes with past effects. Alluding to American Poet Ezra Pound, the authors then claimed that "reality" in the twentieth century, had meant "make the new." This pronouncement also referenced the physicists' observation that, ontologically, the only material bond between two things is resonance. In other words, not linear, sequential, or reciprocal action-reaction, but the interplay amongst many factors bouncing off of one another shapes a new outcome or reality. As described in the article, causality can thus be understood, at least epistemologically, as a "process pattern," whereby causes are perceived ("exposed by discovery") or conceived ("imposed by invention") (10). The effect is that that changes in ground, foreshadow

199 SHOT is the Society for the History of Technology.

understandings of changes to historical subjects, or *figures*. If this figure/ground understanding of causality is taken to its logical extent, effects often proceed causes. Once this observation has been acknowledged, what could and should be explored, the authors advanced, was the process by which effects become causes. Borrowing from quantum physics, they called this process "complementarity." In the context of communication, studying such a process that results from an interplay of factors is done by studying media, by understanding the ground (or environment) from which changes can emerge (11). By extension, this means that knowledge and awareness of media allow insight into the epistemology of causality in the electric world: "Our studies of media as environments that alter patterns of perception and sensibility are intended to develop awareness of the process by which 'new things' come to be regarded as 'facts' " (14).

History in the Electric World – When History Becomes Mythic the Historian Must Become an Artist

McLuhan and Nevitt's analysis of causality can be extended to illustrate a number of consequences for historiographic understanding. These consequences can be thought of as three historiographic illusions: the illusion of a fixed historical subject, the illusion of historical continuity, and the illusion of historical objectivity. First and foremost, rendering certain abstract concepts like technology (figures) into visual metaphors as the fixed cause of past realities is misleading. Conventional historical meaning-making involves the reversal of mechanical causality, meaning {effect→cause} as opposed to {cause→effect}, which can be summed up by the phrase "concepts always follow percepts" (16). This is because, as they argued, "causes are conceived," meaning that the conceptual choices of a conventional historian ("conceptual man") are an attempt, after the fact, to find one or many "missing link[s]" as causes for a certain effect (11). Compounded by the fact that historians are usually unable to explore past realities perceptually, they must trace back from artifacts, stories, and effects to find causes that they can then label. McLuhan critically referred to this process and that which it produced as the "vast apparatus of classifications as a means of controlling and harmonizing experience" whereby "colossal systems of concepts" have become the dominant way to make sense of the past. As a consequence of conceptualizing the past using visual metaphors and goals, the authors claimed that such "conventional ideas" like the concept of technology were an "illusion engendered by the visual bias of visual cultures" (11). Under a visual bias (ground), the mechanical concept of causality dictated that historians place causes at the beginning of sequential connections between separate phenomena (figures). When applied to material change, as in the metanarrative technology drives history, technology serves as the cause of the effect, in this case, history. The fundamental issue with this rendering of the past is that technology itself is a historical concept that was "invented" as a label after the fact to describe real-world changes. Following this logic, assuming that a fixed historical subject called technology causes history is both an illusion and a categorical error.

Second, building on the notion that the ground or environment for communication had been ignored in its role in shaping understandings of causality, McLuhan and Nevitt made the claim that continuity explodes into discontinuity in an electric environment. In other words, the perceived connections between cause and effect can no longer be observed. As a consequence of the acceleration, via electric networks, of a culture that had been based on visually structured components, "history becomes mythic" (2). As the authors noted, a reversal from historical into mythical occurs because "time compression and juxtaposition of events past, present and future merge into electric *nowness.*" In an electric environment, any single point of view can serve as a "multiple and iconic re-presentation" of the past. The "story space" in historical accounts is discontinuous, meaning there is no clear, coherent structure. Historical understanding is caught in a perpetual sprint from moment to moment. The past is illusorily made cohesive (unified) and coherently whole (rendered understandable) through ahistorical concepts or images that are instantly transferrable and transformable. The narrative technology drives history can thus be said to feign continuity of both structure and action in the past.

Third, acknowledgment of the missing ground of visual culture also infers that historical truth itself may not be what it seems; or as the authors claimed, scientific objectivity is revealed as an "optical illusion of truth" (16–17). This statement hints at a critique of the correspondence theory of truth which, broadly understood, embraces the notion that facts correspond or relate to an ontological reality. The authors rejected such an understanding, stating that "truth is never a label; it is not something we match. Truth is something we make with all our senses in a conscious process of remaking the world as the world remakes us physically, psychically, and socially" (17). In an electric world, one should not seek to achieve objectivity by "matching new observations with old concepts," but rather to understand the world through an analysis of communication, meaning various media and their environments (17).

Ultimately, the inference drawn by McLuhan and Nevitt with respect to causality in an electric world is that scholars can serve themselves and their audiences better by striving to balance the practices of a conceptual historian with that of the perceptive artist. This means that they should attempt to displace conventional percepts and achieve understanding as opposed to merely espousing value judgements from the "rear-view mirror." Acting as an artist in this broad sense, scholars can anticipate effects and prevent causes, or as the authors concluded, "we can now anticipate the effects – both services and disservices – before allowing the causes to develop. We no longer merely choose options, we make them" (18).

The Critical "Mumford Manuscript"

The notion that McLuhan might be considered a technological determinist can now be viewed within proper context and therefore dismissed. According to the evidence laid out in the previous sections, it seems rather ironic that the Canadian has been accused of determinism and linear causality when he advocated for an understanding of change that included networks of relations – the galaxy metaphor – and a type of formal causality that has been described as an early version of emergence theory.[200] While McLuhan's writings certainly contain deterministic overtones, this chapter has argued that he understood technology as part of a co-created environment,[201] that non-linearly[202] affects human perception and communication within a historically contingent framework. Building on this interpretation of McLuhan's contribution to discourses of technology and an understanding of the relations between technology, society, and the past, two final, related questions follow; one rhetorical and one central to the current chapter. The first is: have Mumford and other scholars projected a common preoccupation with what Aristotle referred to as efficient cause[203] onto McLuhan's provocatively formulated and exaggerated statements? If this first question is answered in the positive, meaning that McLuhan has been misunderstood as a technological determinist, the second question then becomes: which aspects of McLuhan's critique of technology best articulate a change in discourse to evaluate technological change?

In an attempt to answer this fundamental question, the following section returns the attention to the short manuscript mentioned at the outset of this chapter. It is important to recall that McLuhan never had the manuscript published and denied having a theory of technology. His son Eric, having worked with his father

200 See Chapter 4 of Robert Logan's book *McLuhan Misunderstood* where he argues that "the simplest way to understand McLuhan's position vis-à-vis causality and determinism is to recognize that McLuhan was basically foreshadowing emergence theory if not in name at least in spirit" (Logan 103).
201 Summarized under the "environmental thesis" this is the idea that media are "not the bridges between man and nature: they are nature" (Lister 89).
202 As Lister points out in his section on criticisms of McLuhan's notion of causality: "McLuhan's conception of the causality employed in electronic technologies is non-linear" (335).
203 Aristotle understood efficient cause to signify when materials assume a specific form as caused by an agent.

around the date printed on the last page of the manuscript (28 February 1973), was unaware of its existence. This is because, although he "had been compiling a bibliography of [his] father's written works published and non-published," he "attest[ed]" that he had "never before see a copy of this essay." Moreover, he was "certain that it [had] never been published anywhere" and added anecdotally that, "in 1970, [he] had been working with [his] father as his assistant for at least five years and recall [ed] his reading Mumford's *The Pentagon of Power*, but [he did] not remember his writing this particular piece."[204]

At a minimum, the manuscript's relevance to McLuhan's scholarship should be measured in its ability to crystalize his approach to technology against a major figure in the field. As a direct response to Mumford's criticism in *The Pentagon of Power*, the document is more relevant to demonstrating a change in discourse than any of the other writings analyzed in this chapter. Beyond that, it also manages to retain several general themes that McLuhan used in other writings to fortify his approach to the study of technological change. At maximum relevance, the manuscript should be thought to constitute the pillars of the framework for his own interpretation of a "metamorphic communication theory" as well as a justification for his critique of technology as a shaper and constrainer of human sensibilities and awareness ("Mumford Manuscript" 13). The four themes around which McLuhan's critique of Mumford centered are a concern for form over content, uncovering of an unconscious visual bias, the interplay of figure/ground as a theory of change, and the importance of the awareness of artists.

Concern for Form over Content

Reviewing *The Pentagon of Power*, McLuhan wrote that "Dr. Mumford"[205] spends a lot of time searching into the "crystal ball of the Megamachine" only to discover that it "contains no message; it is the message" ("Mumford Manuscript" 2). Based on this statement, McLuhan revealed his special concern for the Megamachine as a form over the content of Mumford's description. Given the assumption that the "user of any medium is the content" of that medium, following McLuhan's logic, the person inside the Megamachine is the content of the Megamachine (*Letters of Marshall McLuhan* 448). As a consequence of McLuhan's discontinuous historiography of communication, he held a skeptical view of scholars that emphasized

---

204 From the author's personal correspondence with Eric McLuhan from 16 April 2018.

205 It is noteworthy to point out that, in a sign of respect, McLuhan referred to Lewis Mumford as "Dr. Mumford" throughout the manuscript.

continuities through time. McLuhan characterized as “tentative and whimsical” Mumford’s argument in the book that human society was “flipping from animal to human and then into the magnetic-organic-unhuman again” (“Mumford Manuscript” 2). This interpretation of the past was too broad and “lineal” for McLuhan, who instead found “joy in ordinary experience and consciousness” in “the power of the mind to discern pattern and form amidst the utmost confusion of data” and “satisfaction” in “understanding the many intelligible relations and unexpected forms that the physical extensions of our own physical beings have involved us in” (4, 10). For McLuhan, concentrating solely on amassing knowledge and explaining facts was not enough. McLuhan thought it prudent to interpret the norms that develop from the “environment of services and disservices” constituted by technology (5). To the point, McLuhan argued for pattern recognition and discernment of form. He thought that without “play,” without “discontinuity,” “there can only be logic and connection and fixity” (4).

Unconscious Visual Bias

In one of his rhetorically harshest critiques, McLuhan lumped Mumford in with many others as “an unwitting prisoner of visual modalities,” by which he meant scholars that espoused “fixed points of view, dichotomies, classifications, panoramic descriptions and blueprints” (4). As McLuhan argued, Mumford’s unconscious visual bias “betray[ed] him into a pervasive habit of moral quantification, a habit of adding up the losses and the gains of historical development . . . .” According to McLuhan, this “omnipresent visual orientation” revealed Mumford’s “central problem”: “he [had] no theory of communication or change” and could only evaluate change in “the most vaguely moralistic and descriptive terms” (4, 17). Incorrectly through the “typical illusion of naïve visual bias in human perception,” Mumford assumed that “technology is ‘contained’ within some sort of plenum . . .,” that man occupied this plenum, and that “technologies are added as new content to this plenum” (6). Mumford is thus left to employ “the conventional and merely external means of totaling up losses and gains that accrue from innovation” (14–15).

Theory of Change – Interplay of Ground with Figure

Mumford was not alone in receiving criticism for a lack of an explanation for change. McLuhan wrote that many “visually oriented” Western philosophers had presented “figures minus their grounds” and had been “blind to the central metamorphic role of all human artifacts” (14). McLuhan further lamented that patterns of cause and effect had not been studied as they related to the "figure and ground play of artifacts" by other major scholars; not even by Hegel and Karl Marx, both of whom, according to McLuhan, also did not have a theory of communication.

McLuhan thought that response to theorizing change through the figure/ground relation has been misunderstood: an "approach to literature or philosophy," like that of Harold Innis, "through the ground created by media themselves," was thought to be "superficial, technical-materialistic and inhuman" (13). McLuhan criticized Mumford for ignoring Innis and for lacking a "metamorphic communication theory," which understands that media alter the "psychic lives of their users." In particular, a concern for communication means that the change between "visual and acoustic forms of the clock," but also to the uniformity of the price system before and after the printed book can be reexamined and characterized as processes though which the perceptual lives of their users were altered.

Importance of the Awareness of Artists

Based on his writings in *The Pentagon of Power*, Mumford was considered by McLuhan to be a typical historian, not "concerned about the effects of technology on human sensibility and awareness" and "content to generalize about 'those who created the mechanical world picture'" (16).[206] To McLuhan's mind, because Mumford lacked or had neglected the grounds for change in the past, the American had been unable to emphasize that "the heightening and isolation of the visual power was the indispensable condition for the beginning of quantified modern science . . ." (16). Such recognition would have allowed Mumford to understand that "only specialized and visual forms of space" allow for "division and sub-division and quantification."

What was needed, according to McLuhan, was for historians to become more like artists because they could help narrate the past figure/ground relation. McLuhan believed this role was important because "without the artist, men lose their figure-ground relation to their own machines. They merge and become their machines" (9). Much like artists sensing the world in new ways, McLuhan understood the interaction of society with new technologies as constituting new worlds. He acknowledged that society was mostly unaware of the effects of technologies and that one could be "manipulated" into thinking that using technologies consti-

---

206 In contrast, McLuhan praises Siegfried Giedion for being the "first technological student to give adequate and particularized attention to the changes in human sensibility relating to the production and use of everyday objects" ("Mumford Manuscript" 16).

tuted a form of freedom: the "unawareness of the sensory modes and psychological states subliminally communicated by the unrestrained environments released by technology" had been the case for "most people throughout recorded history" and "men have always been manipulated by their own inventions and called it 'freedom'" (10–11). However, for McLuhan, studying technology and writing about its past should not only be concerned with recanting the process by which man became mechanical; it should also focus on awareness and emancipation from the shaping factors of the media environment: "awareness of the sensory and perceptual effects of diverse technologies can make possible a humane and modest existence for those who seek it." Moreover, understanding [media] created "an available means of liberation and deliverance" and not "commitment" to a moral stance or static perspective or worldview. Ultimately, this is the major difference between Mumford and McLuhan: Mumford invites his readers to commit to the moral imperative of reorienting technics, while McLuhan summons the percepts of the artist to understand the dynamic effects of media in order to deploy countermeasures.

Chapter 6
Creating a New Technological Narrative

The aim of the real eutopian is the culture of his environment,
. . . not the exploitation, of some other person's environment.
Lewis Mumford, *The Story of Utopias*

At the heart of this book lies an observation about the historiography of technology during the twentieth century: frameworks to explain material change shifted from those that viewed technology as a potential determinant of historical change to those that understood technology, technologies, and technological systems as embedded within a dynamic historical and cultural setting. Thanks to the work of many scholars of technology, it is now easy to recognize a broader and more abstract question that has been superseded by this shift in understanding from necessity to contingency, namely, the degree to which technology determines history, or technological determinism.

The first chapter of the book established that deterministic and progressive conceptions of what is now referred to as technology were particularly dominant in the nineteenth century. In similar fashion, the early twentieth century was demonstrated to have witnessed a continuation and, in certain corners, a strengthening of this conceptualization of technology. To this latter point, a reductive and instrumental concept of technology was used by a prominent historian, Beard, and two sociologists, Parsons and Ogburn, to explain changes in society. A technical and social order and control like those advocated by Taylorism and Fordism[207] can also serve as a prime example of how an instrumental concept of technology took command of narratives surrounding the "mechanization and systemization" of America during this time (Hughes, "Machines, Megamachines, and Systems" 111).

The middle chapters of the book (two, three, four, five) attempted to trace an opposing trajectory in the historiography of technology through several North American writers. Their works described and theorized a subject of historical inquiry that was conceptually different, with Mumford preferring even a different name for the subject matter – technics. What is more, this literary group of technology's storytellers used context and contingency to interpret the social effects of technological change. By reconstructing an intellectual lineage that connects Veblen to Chase, Mumford, and, to a lesser extent, to Marshall McLuhan, the pre-

207 Consistent with Thomas P. Hughes' definitions, Taylorism can broadly be understood as scientific management, while Fordism can be shorthand for mass production (Hughes, "Machines, Megamachines, and Systems" 110).

https://doi.org/10.1515/9783111396477-007

ceding pages make the claim that these historian authors have contributed in foundational ways to the historiography of technology in North America. To underscore this point, Chase, Mumford, and McLuhan, public intellectuals in their own time, all received an education and wrote most of their work before any formal organization of the history of technology was established or any journal specific to the history of technology had been published. The decision to treat these writers as a group was put forward as evidence that a shift in frameworks from necessity to contingency in the historiography of technology had already begun to occur in the early twentieth century. Despite attempts to combat frameworks of necessity as far back as the 1920s, technological determinism has remained very much alive in the popular imagination. Did the efforts of these and other authors fail to materialize in more debate around narratives of technological determinism? It is fair to ask this question in light of the marginal role that Veblen, Chase, and McLuhan have played in discourses about the beginnings of the history of technology in North America.

This final chapter of the book thus connects the authors in the current study with the emergence of formal and institutional inquiry into technology as a historiographic subject. This move shows how the history of technology was taken up by later groups of scholars that increasingly rejected frameworks of necessity and embraced frameworks of contingency. An account of the latest developments in the field would be far too detailed for this purpose. Instead, several points of emphasis are offered in an attempt to bring the authors in this study into conversation with the historiography of scholars that studied the history of technology from 1960 to 1980. What follows is an effort to capture a few ways in which scholars of technology subsequently moved beyond frameworks of necessity, narratives of progress, and technological determinism at the end of the twentieth century.

In order to emphasize perhaps the biggest change of the last century in the formal and institutional inquiry into the history of technology, a short summary is provided of the beginnings of the Society for the History of Technology (SHOT) and its journal, *Technology and Culture*. Connections, or lack of connection, between the authors in this study and the founding of SHOT and any publications in *Technology and Culture* are indicated. A second section describes the type of historiography that is characteristic of SHOT members in their early days. A third section returns to a claim first made in chapter one where Chase, Mumford, and McLuhan were referred to as "prophets of contingency." This leads to probing whether the focus of this group of authors was conceptually neglected, or not adequately acknowledged, at the beginnings of SHOT. The fourth and fifth sections of this chapter present specific examples of how frameworks of necessity have subsequently been broken by other scholars. A first example involves a variant of "soft determinism" found on *Technology and Culture*'s pages. Citing this notion provides an example of how SHOT members have attempted to find an antidote to the ambiguities in understanding surrounding deterministic narratives.

The next section briefly elaborates on the position of a classic Marxist text in the field before listing other subsequent shifts in who and what is covered in accounts of the technological past. It should be clear that the field now has a wide-open definition of the people and things that can be studied within what is referred to as the "narrative framework" (Staudenmaier, "SHOT at 50" 630). In some final thoughts on the shift from frameworks of necessity to frameworks of contingency, the discussion returns to Veblen and possible traces of his effect on discourses of technology. The penultimate section points out a small, but substantial language matter, and offers a final statement on McLuhan's contribution to discourses on technology. The chapter concludes with a short reflection on what has been neglected in discourses about technology outside the academy (not necessarily in practices among academics), and with observations that are relevant in the twenty-first century.

Formal and Institutional Inquiry into Technology as a Historiographic Subject

In the late 1950s, a group of scholars that included Mel Kranzberg, who had mainly been a scholar of French history up until that time, approached a leader in the History of Science Society (HHS) about having the history of technology included within the scope of the larger and established HSS (Staudenmaier, *Technology's Storytellers* 1).[208] The group's proposal was rejected, which led directly to the founding of the Society for the History of Technology in 1958 (223). Thereafter, academic discourses on technology were consolidated in a group of SHOT contributing scholars and institutionally separated from those on science. Thus, the beginnings of continuity in an academic specialty can be traced to the founding of SHOT and in work for the society's journal, *Technology and Culture*, which provided the scholarly space to engage in research and communication that helped forge a collective and professional identity.

By the mid-1960s, the society and its journal had begun to show signs of stability, both institutionally and in terms of scholarly identity (Staudenmaier, *Tech-*

208 Among other detailed histories of this initial period, see also Robert Post's essay "Chance and Contingency: Putting Mel Kranzberg in Context," which considers Kranzberg's career itself as a study in contingency, and another essay "Back at the Start: History and Technology and Culture," which details some of the vital exchanges in those first years.

nology's Storytellers 2). Kranzberg's endeavor[209] to create "an institutional matrix to facilitate a new understanding of the past" could be characterized as successful (Post, "Back at the Start" 963). That is, until a precarious financial situation, particularly between 1964–1970, when the financial support of Edgar Kaufmann Jr. and the Kaufmann Foundation was required to pay off its budgetary deficits (Staudenmaier, *Technology's Storytellers* 222). Without the monetary infusion by a wealthy man sympathetic to its cause, discourse on technology and culture would have fallen silent during the society's first decade of existence, at least formally and institutionally in the form of SHOT (1–2). It did not, and the journal gained in reputation in its early years, partly due to the prominence of the journal's authors (224). These "big names" were almost exclusively male and English-speaking from the United States (Kranzberg in Staudenmaier 3–4). This fact alone lends itself to clarify to what extent the authors in this study were among its founders or if they published articles in *Technology and Culture.*

The later careers of Mumford and McLuhan coincided with SHOT's early years, but none of the three main figures in this book are considered to be among the society's creators. Of the three, Mumford has been the most influential among SHOT's actual founders, and for that matter, subsequent generations of professional historians of technology. This is mainly because his first work on technics provided a synthesis of technological change organized into distinct historical eras – a context for the emergence of seemingly diverse technologies. Indeed, Mumford "actively supported" Kranzberg's editorial work in the early days, even once recommending in correspondence that SHOT's journal be called *Technics and Culture* (Staudenmaier, "Recent Trends" 178 Footnote 2 and Schatzberg, "Technology" 150).[210] His connection to SHOT also includes four original articles and a book review that were published in *Technology and Culture.*[211] McLuhan has been of less interest to professional historians of technology, partly because he focused primar-

209 Robert Post characterized Kranzberg's grand strategy as an insistence on the name "technology and culture" to bring disciplines together, and his efforts to win over interlocutors to a new way of understanding the connection between the two (Post 963).

210 In a January 1958 letter to Kranzberg, Mumford explained why he preferred technics to technology in the journal's name: "[. . .] Technics is an accurate and useful term for denoting the whole field, and though technology has been used for this purpose in English since 1859, it would be better reserved for the systematic investigation of methods and processes" (as quoted in Schatzberg "Technology" 150).

211 All of Lewis Mumford's appearances in *Technology and Culture* were in the 1960s:

(1) Book review: "Tools and the Man." vol. 1, no. 4, 1960, pp. 320–334.
(2) Original article: "History: Neglected Clue to Technological Change." vol. 2, no. 3, 1961, pp. 230–236.
(3) Original article: "Authoritarian and Democratic Technics." vol. 5, no. 1, 1964, pp. 1–8.

ily on "communication," and partly because some have tended to categorize him as a determinist. McLuhan published one article and two book reviews in *Technology and Culture*. He also had two "letters to the editor" published, which centered around feedback to his "Laws of Media."[212] Chase did not publish in SHOT's journal and is least likely to be referred to by authors writing in *Technology and Culture*, if at all. This is likely because he was primarily trained as an economist, addressed the topic of the Machine Age explicitly in one book, wrote in a more eclectic, journalistic style, and was not overly concerned with the historical past, having focused many of his other books on the trends and policies that would play out in the more immediate future of his own era.[213]

Writers of Contextualist History

The formal organization (SHOT), the journal (*Technology and Culture*), and its annual meetings were novel with regard to the academic study of the history of technology. These institutions certainly helped professionalize and foster new research, but did they fully constitute a "new understanding of the past," as quoted above? And if so, what was characteristic of the new approach? A standard reference work that traces and then characterizes the first two decades of the field's development is historian of technology John M. Staudenmaier's 1985 book *Technology's Storytellers*. Staudenmaier observed that, concerning methodological styles, histories of technology before SHOT was founded tended to employ what he called "internalist" approaches (*Technology's Storytellers* 10). This internalist methodology involved questions of design, or "the specific technical characteristics and constraints of a particular technology under study" (Smith and Reber 134). A second approach used prior to the founding of SHOT involved nonhistorical analyses, by which he meant

(4) Original article: "Man the Finder." vol. 6, no. 3, 1965, pp. 375–381.
(5) Original article: "Technics and the Nature of Man." vol. 7, no. 3, 1966, pp. 303–317.

212 All of Marshall McLuhan's appearances in *Technology and Culture* were in the 1970s:
(1) Original article: "The Argument: Causality in the Electric World." vol. 14, no. 1, 1973, pp. 1–18.
(2) Letter to the Editor: "McLuhan's Laws of the Media." vol. 16, no. 1, 1975, pp. 74–78.
(3) Letter to the Editor: "Misunderstanding the Media's Laws." vol. 17, no. 2, 1976, p. 263.
(4) Book review: "Television: Technology and Cultural Form." vol. 19, no. 2, 1978, pp. 259–261.
(5) Book review: "The Dialectic of Ideology and Technology." vol. 20, no. 1, 1979, pp. 244–246.

213 Chase understood the types of people, institutions, economic activity, and systems that were harmful and in need of reform, reorientation, or even dissolution. For example, in his book *A New Deal* (1932), Chase advocated for a "third road" of action that would allow industry [the production and distribution of technologies] to be managed in the public interest. This strategy represented an attempt to "dissolve capitalism with a minimum of governmental interference" (241).

research based on sociology, economics, and political science (Staudenmaier, *Technology's Storytellers* 11). Attempting to steer publications in a different direction and "concerned about the limits of internalism," as Smith and Reber have also noted, the founders of SHOT attempted to foster an "alternative, contextual" methodological approach[214] to the study of technology during the 1960s and 1970s. In principle, this approach was an attempt to understand the relationship between design and the economic, social, and political contexts in which a particular design emerged and developed. (Smith and Reber 134). Increasing use of contextualism in the methodology of technology studies, it was hoped, might lead to a decrease in the acceptance amongst historians of the inevitability of technological progress and the ability to transcend the internalist approach. As Staudenmaier expressed, this was also done in an effort to commit to a method that could create a synthesis out of internal design and societal context:

> By their choice of *Technology and Culture* for the journal title the founders [of SHOT] announced their desire to create a new style of history, which Kranzberg and others would begin to call "contextual history." In 1958 aspects of the new approach could already been seen in works by scholars such as Louis Hunter, George R. Taylor, Matthew Josephson, I. B. Holly, Lynn White Jr., Abbott Payson Usher, and Mumford himself. What was lacking was an explicit commitment by an identifiable community of scholars to a method that respected both the subtleties of technical design and the complexities of the historical ambiance. It would not serve simply to juxtapose design and context. Only by integrating both dimensions would the new style begin to interpret technical designs as part of history's single woven fabric. The creation of a new historical language is challenging by definition. Contextualism was doubly difficult since it demanded technical as well as historical expertise (*Technology's Storytellers* 163).

Building from this quote, not only the institutions, but also having propagated a defined purpose – the commitment to the contextual style by a discernable group of scholars – was what made SHOT different. Not necessarily the contextualist style itself. The books and authors' names invoked as suggested antecedents to this historiographical style, at least according to certain historians of technology, have resulted in some common answers. In 1989, when Merritt Roe Smith and Steven C. Reber could still call the history of technology "a relatively new field of inquiry," they acknowledged the historical study of technology in the first half of the twentieth century by mentioning the contributions of a handful of scholars like Louis C. Hunter, Joseph W. Roe, Abbott P. Usher, Lynn White Jr., and Lewis

214 The contextual style is one in which a piece of writing discusses "the functional design of the given artefact(s) and also discusses some aspect(s) of the ambiance in which the artefact(s) exist(s)" (Staudenmaier, *Technology's Storytellers* 205).

Mumford. But these men, they contended, worked in isolation, and "lacked any shared notion of professional identity." (Smith and Reber 133).

Staudenmaier provided hard evidence that Mumford was unique in serving as a role model to at least one of SHOT's founding members with regard to writing contextual technological history. Citing correspondence from Kranzberg to himself, Staudenmaier revealed the level of admiration Kranzberg had for Mumford's *Technics and Civilization*:

> Lewis Mumford's *Technics and Civilization* had such a *great* impact on me; I read it as an undergraduate at Amherst when it was first published back in 1934, and it has stuck with me ever since – and I continue to assign it to my classes because it is still a mind-blowing experience for them! (Kranzberg to Staudenmaier, 4 March 1983) (*Technology's Storytellers* 242–243).

Due to responses like the above, Staudenmaier specifically identified the impact Abbott P. Usher's and Lewis Mumford's books had had on SHOT's methodological choice and on the academic field of the history of technology. He designated their two classic texts, *A History of Mechanical Inventions* (1929) and *Technics and Civilization* (1934), as "forerunners" of the contextual style.

The same texts were likewise singled out as significant to the historiography of technology by historian Arthur P. Molella. In a book chapter entitled "The First Generation: Usher, Mumford, and Giedion," Molella explored how, during the late 1920s and early 1930s, three historians' studies[215] expanded from the traditional "internal issues of engineering and invention" in an "inert world" with "artificial objects" to address greater questions surrounding the development of technology (Molella 88–89). As the title of his article suggests, Molella framed the authors of these works as a first generation of scholars that offered a new, more societal approach to a much longer history of invention. In other words, he argued that they were interested in a history of technology as a kind of social history that synthesized the knowledge of different academic disciplines. Importantly, as Molella claims, they were some of the first authors to subject mechanized society to "serious historical scrutiny," which made their books "landmark studies" that greatly influenced the development of the history of technology as an academic field (88). This later point is certainly true, yet as the current study has demonstrated, there

215 The article explored Abbot Payson Usher's *A History of Mechanical Inventions* (1929), Lewis Mumford's *Technics and Civilization* (1934), and Sigfried Giedion's *Mechanization Takes Command* (1948). Molella's article suggested that, despite the span of years between their publication dates, all three authors were working around the same time and attempted to respond to the "shortcomings of technological society" (Molella 88). Obviously, the crossover with the current study is Lewis Mumford.

were other authors writing about the tenuousness of technology in social context during the early-to-mid twentieth century. Might their works be considered contextualist in style and thus a link to others both past and present?

Foretellers of Technology's Impact

In the introduction, Chase, Mumford, and McLuhan were referred to as prophets of contingency, a narrative choice that can now be more fully understood. First and foremost, it seems appropriate to use the term prophet as a metaphor to call attention to their clear and vivid awareness of the downsides of technology that others had ignored or discounted. The word prophet is often misused only to signify the supernatural. But here it should be understood to summarize the authors' contributions to the historical critique of technology because, like prophets delivering a message, they attempted to convey the idea that technology's further development was contingent on humans' relation to it. Combining both emerging discourses surrounding technology and a delusion with necessary links, like those between technological, media, and economic systems, these men intended their ideas to serve as a form of social critique. In a broad sense, their histories shared a narrative framework for technology that held the potential to trigger political change and that envisioned greater independence from dominant technological orders.

Prophets are not necessarily forerunners, as Staudenmaier characterized Mumford's more contextual approach to historiography, but foretellers; people who, through moral insight and exceptional powers of expression, can speak to possible future implications of their subject of inquiry (*Technology's Storytellers* 12). As established in the individual chapters, Chase, Mumford, and McLuhan all wrote about the impact of the dominant industrial (technological) order, both in the past, and in their times. In as such, they attempted to warn of the dangers inherent in frameworks of necessity and a blind allegiance to overly progressive narratives of technology and its past as conveyed by those with a vested interest, not in prophets, but in profit.

Prophets also have followers and lineages. If a historian of technology like Molella was prepared to include Mumford in a so-called first generation of contextualist historians of technology, it is safe to assume that he, from a distance of two generations,[216] viewed himself and others like him to be the intellectual descendants of these scholars and their works. Beyond that, Molella's use of the word generation suggests both a temporal (as opposed to a semantic) grouping of the authors as well as an implied continuity with the present. If the latter is true, based on the evidence presented in the preceding chapters, this book has argued for Chase and McLuhan to be included with Mumford as authors that have engaged in contextualist historiography and historical narrative that employed frameworks of contingency.

As authors, Chase and McLuhan have been marginalized or forgotten as part of a conversation about early discourses on the history of technology. On the contrary, the lack of attention to the subject matter they were interested in did not go fully unnoticed. At the end of *Technology's Storytellers*, Staudenmaier considered questions that were rarely addressed on the pages of *Technology and Culture* in its first two decades of its existence. As Staudenmaier offered, "culture conflict in technology transfer," or consideration of the cultural dimensions in transfers of technology had been generally overlooked (*Technology's Storytellers* 177–178). It stands to argue whether McLuhan's work on changes in media or the cultural changes wrought by the introduction of new media, in this context electronic media, could have alleviated this omission. Another theme that had received "peripheral status" in *Technology and Culture* articles in its early years was a critique of capitalism (179). Due at least in part to the influence of Veblen, Chase's and Mumford's early texts[217] openly criticize industrial production and industrial values, and the impact they had had on workers, consumers, environments, and even on aesthetics. Stuart Chase wrote about industrial waste and surveyed the effects of technology in society, and was an early advocate for consumer protection. Mumford included in his account a prescriptive list of recommendations to align technics with societal and cultural values that promoted a culture of life.

To his credit, Staudenmaier was well aware of the many thematic shortcomings of the field at an early stage. For example, he critically understood that individuals, groups, or institutions who were affected by the design of a technology or the "rigidities and limitations inherent in the technical dynamic," made up a so-called "impact constituency," and were often ignored in treatments of technology (198). In fact, in the preface to the 1989 paperback edition to *Storytellers*, Staudenmaier strongly emphasized his hope that including an impact constituency as

216 Regarding the impact that Usher, Mumford, and Giedeon had on successive historians of technology, Molella wrote that, "from the distance of two generations, these three seemingly spontaneous products still stand out as singular creations" (Molella 88).
217 In particular, Chase's *The Tragedy of Waste* (1925) and Mumford's *Technics and Civilization* (1934).

an “essential dimension of technological context” would foment serious discussion of the ways in which “successful technologies oppress even as they bless” (Staudenmaier xxiv). This invisibility in early academic articles on technology could certainly have been remedied by inclusion of the authors in this book. In fact, Chase, Mumford, and McLuhan could be considered authors who made it their life’s work to call attention to the cultural aspects of technology, the impact of technology, technologies, media, and the technological/industrial order and values in society. And this is precisely why they were treated as a group in this study. But looking beyond this group, it is important to name a few other attempts to break the framework of necessity that were articulated at the end or after the careers of Chase, Mumford, and McLuhan.

Technological Momentum

Staudenmaier argued that SHOT’s main task had been to develop a “contextual interpretation of the relationship between technology and culture” to combat what he called the “progress talk” embedded in frameworks of necessity (*Technology’s Storytellers* 148). So how did SHOT historians contribute to resolving determinist narratives of technology during their first two decades of publication? One way in which this was accomplished was embracing a soft version of determinism that could exist within a broader framework of contingency. Although rarely invoked explicitly by name during the first two decades of SHOT, historians writing in *Technology and Culture* frequently wrote about technological change as a kind of “momentum.” In *Storytellers*, Staudenmaier reviewed this concept, first used by historian of technology Thomas P. Hughes in 1969, as a metaphor to understand the complexities of the relationship between technology and culture, and to interpret the impact of technological change. But because a consensus in language had yet to be achieved about the relationship between technology and culture, Staudenmaier included in the discussion about momentum the “enduring nature of (1) existing technical concepts (2) existing technological artifacts (3) governmental policy (4) financial vested interests (5) technological enthusiasm (6) cultural values” (*Technology’s Storytellers* 155–160). In a more recent publication, Hughes described the metaphor of technological momentum as follows:

> Technological systems, even after prolonged growth and consolidation, do not become autonomous; they acquire momentum. They have a mass of technical and organizational components; they possess direction, or goals; and they display a rate of growth suggesting velocity. A high level of momentum often causes observers to assume that a technological system has become autonomous. Mature systems have a quality that is analogous, therefore, to inertia of motion. The large mass of a technological system arises especially from the or-

> ganizations and people committed by various interests to the system. Manufacturing corporations, public and private utilities, industrial and government research laboratories, investment and banking houses, sections of technical and scientific societies, departments in educational institutions, and regulatory bodies add greatly to the momentum of modern electric light and power systems. Inventors, engineers, scientists, managers, owners, investors, financiers, civil servants, and politicians often have vested interests in the growth and durability of a system. Communities of practitioners, especially engineers maintaining a tradition of technological practice, sometimes avoid deskilling by furthering a system in which they have a stake. [. . .] Concepts related to momentum include vested interests, fixed assets, and sunk costs ("The Evolution of Large Technological Systems" 70).

Put simply, a technological system can be both an effect and a cause. Younger systems are more contingent on social and historical factors, while older systems tend to be more dominant in shaping outcomes. This is not to suggest that older systems with political and economic power cannot be changed. Rather, it may be necessary to apply a greater amount of pressure to its various components to force a change.

As a metaphor, technological momentum attempts to take both structural factors and contingent events into account. Clearly, all metaphors should be subject to criticism and scrutiny. But the strength of a metaphor can be measured in its ability to capture complexity without being overly simplistic. Hughes argued that momentum is a useful concept because it "does not contradict the doctrine of social construction of technology,[218] and it does not support the erroneous belief in technological determinism" ("The Evolution of Large Technological Systems" 73). In doing so, it attempts to offer a middle ground or compromise between these two positions. But regardless of its relative value in academic discussions about the relationship between technology and culture, technological momentum represents an attempt by a SHOT contributor to break the framework of necessity by embedding technologies and technological systems within a cultural setting, which can both be more or less resistant to change.

Marxist and Other Narrative Frameworks

By framing the narrative of their work, historians shape understanding of what and who matter in an account of the technological past. In this context, it is worth mentioning an older (1977), but still one of the most authoritative, contextualist accounts of the emergence of modern technology in North America during the late nineteenth and early twentieth centuries – David F. Noble's *America by Design*. In the introduction of the book, Noble indicated how the meaning of the

218 For a definition of social constructivism, see the next section.

word "technology" continued to be mystified. The reason for this, he observed, was that when the story of technology is abstracted from an account of actual historical and social processes, the result is a reduced and distorted understanding of how technological change happens. He added that common language habits of talking or writing about technology as if it was a separate entity from society/culture also reinforced a kind of conceptual split that did not exist in practice (Noble xviii).

To break the framework of necessity in common narratives about technological change, Noble adhered to a classic Marxist view in which the "fundamental relationship between society (social relations) and technology (forces of production) is dialectical." Both, he claimed, were different aspects of one process of social production (xix). Critically, for Noble, the term technology included in its meaning more than just "information, logic, and things."[219] Technology also involved people and their "activities in particular social and historical contexts, with particular interests and aims." The process of technology therefore also reflected the larger social order in which it was situated with all its contours and possibilities (xxii).

The contextualized reason why modern technology developed as it did in North America, Noble claimed, was due to the simultaneous rise of corporate social production (corporate capitalism). Noble thus detailed the "capitalist wedding of science to the useful arts" through the "rise of science-based industry, the development of technical education, and the emergence of the professional engineers" (xxiv). By describing in social and historical context how the resulting and conflated understanding of modern technology was put into practice, Noble could demonstrate that America was "designed" in accordance with this particular perception of social production, in this context, scientific management (xxv–xxvi).

In the 1980s, this type of contingent and contextual framing of the social process of technology was fully embraced by a group of scholars that called their research approach the Social Construction of Technology (SCOT). Its starting points were fundamental criticism of technological determinism and the definition of technology as inherently social. A core idea of SCOT holds that technological designs do not succeed solely because they are the best design, and thus defines "technological success as contingent and inherently political" (Staudenmaier, "Recent Trends" 173). According to one of its main contributors, W.E. Bijker, some of SCOT's central concepts are "relevant social group," "interpretive flexibility," "closure," "stabilization," "technological frame," and "inclusion" (Bijker 15522–15527)

219 Noble deems an understanding of technology as inadequate when the "composite of the assimilated scientific knowledge, technical skills, implements, logical habits, material products of people" does not include the people that do and have these things (Noble xxii).

To consider only one of the concepts above, closure, which is the "moment when some group finalizes the design," reveals the degree to which contingency has replaced necessity as a framework for technological narratives.

By the year 2000, many historians of technology considered refutation of frameworks of necessity with complex case studies "no longer an adequate rationale for publication" (Staudenmaier "Recent Trends" 176). More importantly, due to a historical lack of different voices and perspectives, it was fair to ask how representative discourses on technology's past had actually been. For its part, SHOT and its contributors have grown and diversified, as have its language and themes. What has replaced the early field's narrow definition of what counted as technology, when it started and stopped, and who was considered a maker of technology? According to Staudenmaier, these questions were representative of technology's increasingly contested terrain. At the end of the twentieth century, scholars were putting great emphasis on contingency when framing narratives about technology in cultural studies, which included the categories of "gender, race, non-western perspectives, and studies of user redefinition of technologies" (176). Writing new "actors" (particularly women) into narratives about the technological past showed how inherently contingent the definition of technology had become. But the time frame for an account of technology had also changed. Ordinary use of technology, after it has been designed and released to users, was of particular interest to scholars. This is because any serious historical narrative about technology can now only be framed contingently once they "extend out into the tangle of the human condition" (171).

Veblen's Wake

Veblen understood the importance of technology as part of this human condition – the cultural aspects of technology – and thus criticized modern industry and the industrial order of his time. It is an approach not only worth remembering, but there is also value in tracing the wake he and his scholarship left behind. As Schatzberg noted, it is Lewis Mumford who best represents this cultural view of technology: "in many ways, Mumford's technics represented a full-throated appropriation of the German discourse of *Technik*, especially those aspects emphasizing the cultural relevance of *Technik*. In following the German discourse, in particular Sombart, he also remained true to the cultural view embodied in Veblen's concept of technology" (Schatzberg, *Technology* 151). But beyond Mumford, Schatzberg ended his chapter on Veblen's legacy unsure of his precise contribution to the concept of technology. The preceding chapters suggest that also tracing

Veblen's wake through Chase and Innis, and in an indirect manner, McLuhan, can provide some examples.

The private, corporate takeover of technology in North America was neither necessary, nor inevitable. If technology is understood both as a social process of production and an expression of an inclusive human culture, "the handiwork and rightful heritage of all humans," and not as the product of a narrow strand of private interests, a critique of technology must be accompanied by a critique of the conditions in which it has arisen (Schatzberg, *Technology* 15). Through their writings, the authors examined in this book cautioned against blind adherence to the advancement of collective and broad nouns like technology.[220] The opposite or status quo position, they inform, only helps uphold and stabilize frameworks of necessity. Specifically, both Chase and Mumford sympathized with pacifist and egalitarian ideological positions; both argued for the need to stop worker and environmental degradation, war, and to fundamentally reform capitalism to ultimately go beyond the profit-motive and reap more positive and collective benefits from uses of technology. Collectively, these two authors remind their readers of the responsibility to expose and dissolve technologies and technological systems that destruct, degrade, or disregard humans, plants, animals and the environment.

As evident in the previous two chapters, Mumford and McLuhan were extremely insightful authors. But for all their theoretical and historical brilliance, familiarizing oneself with Stuart Chase's ideas can be particularly instructive. Often ignored in scholarship and pejoratively dismissed with the classification of being a technocrat, Chase was arguably more effective in his role as a way-faring popular economist and synthesizer of technology and economy than either of the other two authors. At a time when the Red Deal and Green New Deal[221] have been advanced as viable plans to fundamentally change economic activity in the United States, it is worth reading Chase to understand both the "tenuousness" of capitalism and its industrial order and to read how he was able to imagine a different, more functional relation between concepts of technology and society. It is easy to recognize that the latest and ongoing technological revolutions around alternative energy

220 An example of a contemporary collective noun that evokes similar treatment is the way people talk about "the economy" or "the market" being beneficial forces in society. This blind faith in vacuous verbal labels should elicit a follow-up question, as Chase once posed, namely, "what is an economic system for?" (*New Deal* 1–24).

221 The Green New Deal, proposed by Congresswoman Alexandria Ocasio-Cortez and Senator Ed Markey in 2019, is an attempt to connect social justice movements for housing, health care, free education, and green jobs to the realities of a changing climate and environment. *The Red Deal*, authored by the Red Nation, proposes to go further by placing "anticapitalism and decolonization as central to each social justice struggle, as well as climate change" to radically transform "all social relations between humans and the Earth" (*The Red Deal* 8–9).

sources and information technology could have even more profound impacts on peoples' everyday lives. All the more reason, like Chase attempted through his writing, to push the political imagination as far as it will go to benefit greater society.

What is common to all these scholars is that they seem to value a sense of measure and are concerned about cultural balance. After a new technology is introduced, there is always an adjustment period in which it is open to use and further social construction. Their task, as they saw it, was to observe the inequalities and imbalance in North American culture. Extrapolating and understanding these authors' message, the importance of fostering environmental conditions on balance with basic human needs, may be necessary to help maintain humanity in the face of seemingly uncontrollable technological change.

The Things to Which We Attend

Frameworks of necessity feed into a feeling of helplessness and inaction. They do so by framing the story of technology as a dominant force which determines how one should live and fundamentally shapes how one should relate and interact as a social being. As an effect, they direct possible actors to accept the technological conditioning of life in a particular manner as a given. But the vital message to be taken from frameworks of contingency is that they allow for the emphasis of narratives of technology in, not on, society. This turn of phrase reflects the notion that the selected authors considered technology, following a Veblenian template, to be an aspect of social production within a broader framework of human culture and to interpret technological change as occurring within a particular historical context. These frameworks also apply to the work of Marshall McLuhan, for whom technological change was demonstrated not to be the direct, linear result of changes to technology or media, as many scholars have misread, but rather due to a constellation of cultural factors that developed around changes in forms of communication. The great contribution that McLuhan[222] made to conversations about technology is that he concentrated not on concepts and classifications, but rather on percepts and awareness and, echoing Harold Innis, that he encouraged

222 In a cruel twist of fate for Lewis Mumford, it is Marshall McLuhan who made it into the Deutsches Museum, one of the most prominent sites in Germany for the technological past, present, and future. A large display on McLuhan and his ideas was included in a permanent section called "Image, Script, Codes." Lewis Mumford, who so admired the exhibitions there when he visited in the early 1930s, and who delved into the sources he found in Munich to write *Technics and Civilization*, is not.

his readers to liberate themselves from their medial environments by reflecting on the things to which they attend.

The Invisibility of Technics

The histories presented in the previous chapters reveal a critical awareness of the social relations and human-made environments which have formed in the presence of the political economy in charge of modern technics. Despite being everywhere, many of these relations and entire environments are invisible. Thus, the critical message of this book is that it is up to us to perceive them and shape them toward more favorable outcomes. But how? A first step involves emphasizing a cultural understanding of technology that is greater than products and uses that are designed, produced, and controlled by any one company or corporate group. Put simply, technology, like the word technique, can be understood as a specialized way of doing things. This meaning, coupled with use of the term political economy, strengthens an understanding of the links to social and political factors inherent in an economy that relies so heavily on specialization. Related questions are why, when talking about the economy, has the qualifier "political" been dropped and, when thinking about technology, why have these cultural understandings been largely forgotten?

Let us address the second question first. The short answer was already identified in this chapter: almost full corporate control of industry. Social and cultural factors play little if any role in a political economy whose main goals are mass consumerism, growth, and high profit margins. To be in a position to even perceive technics, one must begin by reframing discourses on technology to think beyond individual devices, products, services, and the technological systems that deliver them. This means that every so often we need to put down our smartphones, look up from our screens, and attend to matters greater than currently implied by the word technology. On a more long-term and ideological level, we need to think beyond the machine. In this regard, it makes good sense to (re)introduce the concepts embedded in Mumford's idea of technics to denote the broader cultural/societal aspects of technology, the social relations and human-made environments, that should be subject to oversight and accountability in any functioning democracy. Democratic refers, following Mumford, to the idea that we should "place what is common to all . . . [people] above that which any organization, institution, or group may claim for itself" ("Authoritarian and Democratic Technics" 1). Examples of these commonalities are both obvious and straightforward: clean air and water, healthy food and environments, universal healthcare, housing for all, quality education and jobs, better and more public transport, and peaceful relations with our neighbors.

Unfortunately, there is a tendency when it comes to discussing technology, to focus on performance and power, and to disregard impact. When pressed to justify the reasons why it is important to pursue certain technologies or improvements to a specific technology, a line is often drawn between autocratic and democratic nations and their systems of control, or lack thereof, like respect for private as opposed to state entrepreneurship, morals, ethics, laws, institutions, and regulations. On one side of this democratic/autocratic divide lie the techno-enthusiasts, who revel in new technologies as a benevolent force that brings improvement to free, open, and democratic societies. On the other side lie the autocrats, who create, copy, and misuse technologies to forge a darkened world that they control. The argument for democracies continuing to pursue technologies of power and performance rests on the assumption that different political cultures and the cultural context in which technologies are assimilated will produce different impacts in the world.

But by and large, at least in the twentieth century, the common pursuit of technologies that focus on performance and power had truly terrible impacts on many cultures and is responsible for cruel environmental injustices all over the planet: these include, but are certainly not limited to the introduction of mechanized warfare, the ever-increasing ability to annihilate other humans, animals, plants, and entire ecosystems, the tragedy of nuclear legacies (in Japan and in many testing zones), a dependency on fossil fuels and mining, an overuse of plastics, a lack of any real solution or accountability for electronic, medical, and chemical waste, and many other aspects of the ongoing ecological crisis. Looking at this list from the perspective of the twenty-first century, one could argue that the conflation of technology with the values of power and performance has blinded us to the larger picture: even some of our "successful" technologies have left many of us and our children more oppressed than blessed.[223] The inability to look beyond technologies we know harm us, and foster a greater ideal, not only for democratic uses of technology, but for uses that allow human culture to flourish is a result of the invisibility of technics.

Why has "political" been almost completely dropped when talking about economies? Among today's political powers (China, Russia, the U.S., and Europe just to name a few), there seems to be little difference with regard to private con-

---

223 Wording similar to the above was used in the introduction to the paperback edition of John M. Staudenmaier's *Technology's Storytellers*. The original language called for the inclusion of an "impact constituency," which would foster more attention to the ways that successful technologies "oppress even as they bless" (xxiv).

trol of technology, hegemonic dominance of progress narratives that favor national, corporate or even personal interests, and frameworks of necessity that empower them. This is especially true in terms of how these cultures have recently developed and employed similar technologies. If a global monoculture around the practices of technology is allowed to persist, we need to face the facts and admit what different cultures around the world all have in common: they all have largely capitalistic economies designed to permit the accumulation of wealth at the very top with little regard to the rest of their populations or the greater environment in which we live, the planet.

Mentioning economies of a certain type is hardly intended to promote rage against amorphic terms like capitalism. Instead, the current inquiry into histories of technology strongly suggests amplifying the idea of democratic technics in everyday language. This will require everyone to first perceive, then pay attention to the societal effects of what is designed and produced. A next step will involve changes to the political economy: to regulate, correct, or dismantle the negative effects of technologies, technological practices, and technological systems. Similarly, attending to the systems of power that govern over what is made and how it is used will help to convince the majority of populations around the world to move ahead on the political goal of attaining an acceptable level of many of the basic commonalities that we all need and deserve. This lofty goal can be achieved if we think beyond the machine and reframe technology.

Epilogue: Impoverished Intelligence and Curbing Our Technological Enthusiasm

I'm trying to elevate small talk to medium talk.
Larry David, *Curb Your Enthusiasm*, Season 8, Episode 6

Understanding Technological Impact

Discourses about technology and potential positive and negative impacts in society remain a site of ongoing dispute between scholars of the humanities, natural scientists, heads of industry, technology experts, advocacy groups, elected representatives, and the greater public. At the time of this book's publication, both the use of existing technologies and the pursuit of certain new technologies seem to be met with growing skepticism. It is a more pessimistic moment than in the 1990s, when the Internet and digital computing were in their commercial infancy. The optimism of the 2000s surrounding social media is also gone. In general, this is because of increased consciousness of the many impending ecological disasters that need to be addressed. But there is also broader awareness of negative effects with regard to technological change, for example, as a result of whistleblower Edward Snowden's actions, books that question the technological status quo like Shoshana Zuboff's *The Age of Surveillance Capitalism,* and a resurgence of labor power. Snowden revealed the details and breadth of U.S. Government malpractices. Zuboff contextualized recent changes in technology into a larger critique of a new and dangerous economic order. And many unions have recently taken up the fight against unfair working contracts and conditions.

To be clear, causes of technological change do not need to be understood in detail. It suffices to understand negative effects when certain technological forms or practices have been introduced. Examples of an acute awareness of negative effects come from the top down like when the Surgeon General of the United States came out with an advisory about "the effects of social media use on youth mental health" (23 May 2023). No longer solely viewed as a seemingly benign tool used to connect to others, social media has met a moment of reckoning:

> the current body of evidence indicates that while social media may have benefits for some children and adolescents, there are ample indicators that social media can also have a profound risk of harm to the mental health and well-being of children and adolescents. At this time, we do not yet have enough evidence to determine if social media is sufficiently safe for children and adolescents. We must acknowledge the growing body of research about potential harms, increase our collective understanding of the risks associated with social

https://doi.org/10.1515/9783111396477-008

> media use, and urgently take action to create safe and healthy digital environments that minimize harm and safeguard children's and adolescents' mental health and well-being during critical stages of development (United States, Department of Health and Human Services, *Social Media and Youth Mental Health: The U.S. Surgeon General's Advisory* 4).

Much grander than the tool metaphor, the Surgeon General's statement acknowledges the "digital environment" that emerges from social media. What is so bad about this environment? Unfortunately, as anyone who has a kid in these ages may know, "adolescents who spent more than 3 hours per day on social media faced double the risk of experiencing poor mental health outcomes including symptoms of depression and anxiety" (6).

The awareness of negative effects also comes from the bottom up, like when a friend of mine noted, quite aptly, that we need to "keep drones and artificial intelligence out of the hands of military guys. You see what is happening now with the way modern warfare is being waged?" Such simple, yet elegant common-sense assertions raise the bar for what we should expect from technology as a cultural practice. We should be asking a host of questions like: do we necessarily have to develop highly sophisticated weapons, including robots and drones, to eliminate people and destroy environments? Which practices in technology sectors have divided us even more? What regulations or rules should governments implement to prevent damage to ourselves and our living environments?

It is easy to identify individual companies or sectors of the economy that should be broken up or scaled down. Similarly, organizing resistance and legal challenges can help fight against unethical practices, improper behaviors, and adversely negative impacts. Yet the problem with regulating "technology" is that many systems, devices, applications, and services have become far too complex for non-experts to understand. Moreover, common language used to describe these so-called technologies can mystify what a technology actually does or what impact it may have. Consider the following example of how specific words change in meaning over a short period of time. A simple landline "phone" from the 1980s served two main purposes – to call and be called. But what is referred to as a (smart)phone in 2024 is a highly advanced device that also tracks and monitors a person's every tap and move. The data collected is subsequently harvested and sold to advertisers and other data companies, all while being used to create sophisticated behavioral prediction models. The word used to signify the so-called technology may render many processes invisible, but the negative impacts remain – loss of privacy and security, and an opening for manipulation. If given the choice over these unwanted practices, is this the future of technology we would like to continue pursuing?

The idea that we can use tools to improve life is small talk. There is a great need to elevate discourses around technology and it is evident that we need a common framework to understand, talk about, and integrate the impacts of many technologies into democratic societies. The furor (often divided into the binary of enthusiasm/fear) around artificial intelligence (AI), which has been worked on under that name since the mid-1950s, is the latest example of a technological pursuit that could use such a common framework. It may seem trivial, but when confronted by the sudden appearance of a supposedly new technology, it is important to first remember that there is a history of technology. This book has certainly argued for a broader framework to understand technology as a cultural practice and maintained that specific histories of technology can be used as a practical guide to the present. When consulted, the history of technology verifies that decades of work and modification have gone into AI systems.

Despite AI having been around for a relatively long time, international cooperation on regulation has yet to get off the ground and the U.S. government just recently started to address the need to mitigate negative effects.[224] On 16 May 2023, the Subcommittee on Privacy, Technology, and the Law held a hearing on the oversight of artificial intelligence. At the hearing, OpenAI CEO Samuel Altman, IBM Chief Privacy & Trust Officer Christina Montgomery, and New York University Professor Emeritus Gary Marcus testified to Congress about the promises and dangers of artificial intelligence. I acknowledge that it is not these people's job to distinguish between words, but Mr. Altman's reinforcement of the tool metaphor while talking about a very complex technological practice like AI during his testimony is regretful (United States, Senate, Judiciary, Subcommittee on Privacy, Technology, and the Law, *Oversight of A.I.*). The metaphor reinforces the pervasive common sense that technology is somehow a thing, external to humans and their culture, which can be used to achieve specific purposes.

While it is misguided to single out one person for use of a common metaphor, Mr. Altman speaking about AI tools is emblematic of a greater issue. Far from being merely a language choice, this framing of the concept demonstrates why critiques of technology are so important. When a broad swath of people starts talking about a complex issue like artificial intelligence, there is a tendency to borrow from common sense used to talk about it. When a simple metaphor like "AI technology is a tool" gets repeated frequently, especially by the specialists themselves, it becomes increasingly difficult to understand technology outside that narrow framework. So, if so-called AI tools are more of a marketing name for particular technological practices, how should one understand this seemingly

224 There have been past attempts to mitigate the negative effects of technological change at a federal level, like the U.S. Government's Office of Technology Assessment (OTA), which was active between 1972 and 1995.

new technology? Is it more than a large-scale prediction machine? What can artificial intelligence actually do and not do? Or more importantly, in order to allow human culture to flourish, what should AI be tasked to do?

Neither Artificial, Nor Fully Intelligent

A look at the two words that make up the name can help demystify the concept of AI and its supposed power. The word artificial seems to indicate an opposition to natural, meaning human, intelligence. But AI can never be artificial as long as humans imagine, program, develop, manufacture, maintain, and use these systems. There is nothing artificial about human labor or human interaction with AI applications. So, what is actually artificial about AI? Let us quickly move on to consider the second word, intelligence. Can applications, systems, and practices be intelligent? Or stated more precisely, what does it actually mean to be intelligent in a human context? In his 1976 book, *Computer Power and Human Reason*, Joseph Weizenbaum, a former professor of computer science at MIT, warned about limited understandings of the term.[225] If intelligence is reduced in meaning to signify only acquire and apply, AI applications and systems seem to be learning better foresight and have an ever-increasing ability to make sense of the feedback they receive. Both qualities are truly impressive and seem superhuman until we reveal that the foresight is most likely statistical probability and the feedback can be solipsistic and lead to an echo chamber.

As Weizenbaum pointed out, such a narrow concept of intelligence lacks the benefits of other modalities of thought like intellect, intuition, and wisdom. In contrast to these modes, the essential character of computing systems, he wrote, is their regularity and obedience to particular sets of commands (41). As transformers of information, it seems odd to render rule acquiring and applying machines the attribute of fully intelligent behavior. This is because intelligence is not absolute and always requires a frame of reference to be measured. Such an impoverished concept of intelligence cannot cover many specific contexts in which a more human intelligence could be applied and for which a less limited concept is required – like addressing the climate crisis on a political level, tackling poverty, caring for a dying relative, or raising children. Ultimately, Weizenbaum concluded that the source of a computer's power, a domain currently dominated by talk about AI, is neither technological nor mathematical, it is us. If we are indeed

225 Lewis Mumford read the entire manuscript and provided Weizenbaum with the wisdom, criticism, and encouragement needed to complete the book (Preface page x).

the medium through which technology gets its power, it follows that questions about what computing systems can or cannot do does not settle the binary argument for or against AI. Once concerns about the future of technology are stated in ethical terms, the limits of AI applicability are revealed. What ought an AI system do? Weizenbaum's answer to this question was no task that demands human intellect or wisdom (227). Put simply, an ethical/human-oriented line of argument exposes one of the technical limitations of current AI systems: they are not intelligent enough to change a baby's diaper.

Beyond a concern for performance and power we need to ask the question: what is the societal value of AI applications and systems, and what is their social purpose? With humans, at least in an ideal situation, our social purpose is to live and thrive in a life-sustaining, peaceful, enriching, and enculturating environment. Machines? Applications? Tools? Devices? Technological systems? Their ultimate social value depends on the whim of their owners, how their aims are conceived, what purposes their makers have programmed into their code, and how they are used. The obvious follow-up question that also should be constantly asked is: do these instrumental purposes fully align with societal values and purposes for all humans, plants, and animals? For private technology firms the question needs to be: what can your company do to make sure human and technological values and purposes align? It is my sense that we should take advantage of this moment of technological skepticism, curb any wave of technological enthusiasm around AI, and drop the tool metaphor. It will help everyone better understand the cultural practices that lead to the creation of environments which can harm us and others.

Authentic Intelligence

Perhaps equally as trivial as acknowledging a history of technology, as Gabriel Winant, a scholar of Pittsburgh's change from technologies of steel to technologies of healthcare has recently reminded us, we may be individuals, but "we live in a society."[226] Building on awareness of this greater dimension demands that we also heed Lewis Mumford's advice and recognize what is wrong with understandings of technology today: the concept "ignores the complex ecological cooperations favorable to human development" (Mumford, "The Automation of Knowledge" 266). Con-

226 The book referred to here is Winant's paperback edition of *The Next Shift – The Fall of Industry and the Rise of Health Care in Rust Belt America* (2023). The article in *N+1* magazine is entitled "We live in a society." See Winant in the works cited section at the back of this book.

sequently, if we intend to halt the uncritical automation of knowledge[227] and disrupt our current trajectory toward an unlivable environment, as a global society, we had better get to work modifying our understanding of technology to create a more democratic and peaceful version. On a more individual level, and in a time when certain applications enable the lines between real and fake to become increasingly blurry, we need quality foresight and diverse feedback about technology more than ever, but we should also remind ourselves of what is authentic in life: the joy of being present with loved ones, the pain of losing close friends and relatives, the experience of making mistakes, and the ability to withdraw and renew our efforts to change, maybe just for the day, into better versions of ourselves.[228]

227 For more critical insights into what is meant by this phrase read "The Automation of Knowledge" by Lewis Mumford.
228 This sentiment was inspired by the following quote from Lewis Mumford: ". . . human modes of behavior are infinitely more complex than automatic systems; and what makes them so is the margin of choice, the freedom to commit and correct errors, to explore unfrequented paths, to incorporate unpredictable accidents with self-defined purposes, to anticipate the unexpected and to "plan the impossible" – all traits that no efficient, automatic system can countenance. In human organisms the whole personality may intervene at any moment to change the tempo or alter the basic pattern" ("The Automation of Knowledge" 264).

Works Cited

Adams, Henry. *The Education of Henry Adams*. Houghton, Mifflin and Co, 1961.
Adams, Henry. *The Degradation of the Democratic Dogma*. Macmillan, 1919.
Adorno, Theodor W. *Prismen – Kulturkritik und Gesellschaft*. Surkamp Verlag. 1955. *Internet Archive*, archive.org/details/prismenkulturkri0000ador/page/n3/mode/2up.
Beard, Charles, editor. *Whither Mankind A Panorama of Modern Civilization*. Longmans, Green and Co., 1928. *Internet Archive*, archive.org/details/whithermankindpa0000bear/page/n5/mode/2up.
Beard, Charles, and Mary Beard. *The Rise of American Civilization*. Macmillan, 1927.
Benesch, Klaus. "Does Technology Drive History? McLuhan, Leo Marx Und Die Materialistische Medientheorie." *McLuhan Neu Lesen: Kritische Analysen Zu Medien Und Kultur Im 21. Jahrhundert*, edited by Derrick De Kerckhove, et al., Transcript, 2008, pp. 95–104.
Ben-Menahem, Yemima. "Historical Necessity and Contingency." *A Companion to the Philosophy of History and Historiography*, edited by Aviezer Tucker, Wiley-Blackwell, 2011, pp. 120–130.
Bijker, Wiebe E. "Technology, Social Construction of." *International Encyclopedia of the Social & Behavioral Sciences*, Edited by Neil J. Smelser and Paul B. Baltes, Pergamon, 2001, pp. 15522–15527.
Bimber, Bruce. "Karl Marx and the Three Faces of Technological Determinism." *Technology Studies*, edited by Rayvon, Fouché, Rayvon, SAGE Publications, 2008, pp. 215–229.
Brey, Philip. "Theorizing Modernity and Technology." *Modernity and Technology*, edited by Thomas Misa, et al., MIT Press, 2003, pp. 33–71.
Brooks, Cleanth, and Robert Penn. *Understanding Fiction*. F.S. Crofts and Co., 1943. *Internet Archive*, archive.org/details/understandingfic0000unse_o3a7/page/n7/mode/2up.
Brooks, Cleanth, and Robert Penn. *Understanding Poetry*. Holt Rinehart and Winston, 1976. *Internet Archive*. archive.org/details/understandingpoe0000unse_x4r3/page/n1/mode/2up.
Carey, James W. "McLuhan and Mumford: The Roots of Modern Media Analysis." *Journal of Communication*, vol. 31, no. 3, September 1981, pp. 162–178.
Chase, Stuart. *A New Deal*. Macmillan, 1932.
Chase, Stuart. "Declaration of Independence." *Harper's Magazine*, vol. 164, December 1931, pp. 29–36.
Chase, Stuart. *Men and Machines*. The MacMillan Company, 1929.
Chase, Stuart. *The Stuart Chase Papers*. Manuscript/Mixed Material. Library of Congress, Washington D.C. 2–3 January 2020.
Chase, Stuart. *Technocracy: An Interpretation*. The John Day Company, 1933.
Chase, Stuart. "The Enemy of Prosperity." *Harper's Magazine*, vol. 161, November 1930, pp. 641–650.
Chase, Stuart. *The Promise of Power*. John Day Company, 1933.
Chase, Stuart. "The Rationalistic View of Life." *The Stuart Chase Papers*. Box 3, Subject File Pages from Journal 1912–13. Library of Congress, Washington D.C. 2 January 2020.
Chase, Stuart. *The Tragedy of Waste*. The MacMillan Company, 1927.
Cowley, Malcom. "Books That Changed Our Minds." *New Republic*, vol. 97, no. 1255, December 1938, pp. 205–207.
Curtis, James M. "McLuhan: The Aesthete as Historian." *Journal of Communication*, vol. 31, no. 3, September 1981, pp. 144–152.
D'Amico, Robert. "Historicism." *A Companion to the Philosophy of History and Historiography*, edited by Aviezer Tucker, Wiley-Blackwell, 2011, pp. 243–252.
Davies, Martin L. "Cognitive Inadequacy: History and the Technocratic Management of an Artificial World." *Rethinking History*, vol. 20, no. 3, July 2016, pp. 1–18.

https://doi.org/10.1515/9783111396477-009

Davies, Martin L. "The Redundancy of History in a Historicized World." *Rethinking History*, vol. 15, no. 3, September 2011, pp. 335–353.
Deibert, Ronald. "Between Essentialism and Constructivism: Harold Innis and World Order Transformations." *The Toronto School of Communication Theory: Interpretations, Extensions, Applications*, University of Toronto Press, 2007, pp. 29–52.
Diggins, John Patrick. *Thorstein Veblen – Theorist of the Leisure Class*. Princeton University Press, 1999.
Etzler, J. A. *The Paradise Within Reach of All Men, Without Labour, By Powers of Nature and Machinery – An Address to Intelligent Men. In Two Parts.* John Brooks, 1836. *Internet Archive*, archive.org/details/paradisewithinre00etzl/page/n5/mode/2up.
Fouché, Rayvon, editor. *Technology Studies*. SAGE Publications, 2008.
Gordon, W. Terrence. *McLuhan: A Guide for the Perplexed*. Continuum, 2010.
Gordon, W. Terrence. "McLuhan's Compass for the Voyage to a World of Electric Words" *The Gutenberg Galaxy: The Making of Typographic Man*. University of Toronto Press, 2011.
Hård, Mikael, and Andrew Jamison, editors. *The Intellectual Appropriation of Technology: Discourses on Modernity, 1900–1939*. MIT Press, 1998.
"Our History." *Canadian International Council*, thecic.org/about/our-history. Accessed 22 July 2018.
"Our History." *University of St. Michael's College*, stmikes.utoronto.ca/about-us/our-history. Accessed 10 November 2018.
Hughes, Thomas P. "Machines, Megamachines, and Systems." *In Context: History and the History of Technology: Essays in Honor of Melvin Kranzberg*, edited by Stephen H. Cutcliffe and Robert C. Post. Lehigh University Press, 1989, pp. 106–119.
Hughes, Thomas P. "The Evolution of Large Technological Systems." *The Social Construction of Technological Systems, Anniversary Edition: New Directions in the Sociology and History of Technology*, edited by Wiebe E. Bijker, et al., MIT Press, 2012.
Hughes, Thomas Parke, and Agatha C. Hughes. *Lewis Mumford: Public Intellectual*. Oxford University Press, 1990.
Innis, Harold. "A Bibliography of Thorstein Veblen." *The Southwestern Political and Social Science Quarterly*, vol. 10, no. 1, June 1929, pp. 1–13.
Innis, Harold. *Changing Concepts of Time*. University of Toronto Press, 1952.
Innis, Harold. *Essays in Canadian Economic History*. University of Toronto Press. 1956.
Innis, Harold. *The Bias of Communication*. University of Toronto Press, 1964.
Jamison, Andrew. "American Anxieties: Technology and the Reshaping of Republican Values." *The Intellectual Appropriation of Technology: Discourses on Modernity, 1900–1939*, Edited by Mikael Hård and Andrew Jamison, MIT Press, 1998, pp. 69–100.
Jordan, Stefan. *Theorien und Methoden der Geschichtswissenschaft*. Schöningh, 2009.
Katz, Elihu. "The Toronto School and Communication Research." *The Toronto School of Communication Theory: Interpretations, Extensions, Applications*, Edited by Rita Watson and Menahem Blondheim. University of Toronto Press, 2007. pp. 1–6.
Kelley, Florence. *Modern Industry in Relation to the Family, Health, Education, Morality*. New York: Longmans, 1914. *Internet Archive*, archive.org/details/modernindustryin00kellrich/mode/2up.
Kirchberger, Ulrike. *Aspekte Deutsch-Britischer Expansion: Die Überseeinteressen Der Deutschen Migranten in Grossbritannien in Der Mitte Des 19. Jahrhunderts*. F. Steiner, 1999.
Kolmer, Lothar. *Geschichtstheorien*. Fink, 2008.
Kroker, Arthur. *Technology and the Canadian Mind: Innis/McLuhan/Grant*. St. Martin's Press, 1985.
Li, Shuxue. *Lewis Mumford: Critic of Culture and Civilization*. Peter Lang, 2009.
Lister, Martin, editor. *New Media: A Critical Introduction*. Routledge, 2009.

Logan, Robert. *McLuhan Misunderstood: Setting the Record Straight*. The Key Publishing House Inc., 2013.

Lorenz, Chris. "Scientific Historiography." *A Companion to the Philosophy of History and Historiography*, edited by Aviezer Tucker, Wiley-Blackwell, 2011, pp. 393–403.

Lorenz, Chris, and Berber Bevernage, editors. *Breaking up Time: Negotiating the Borders between Present, Past and Future*. Vandenhoeck & Ruprecht, 2013.

MacKenzie, Donald A. "Marx and the Machine." *Technology Studies*, edited by Rayvon Fouché, SAGE Publications, 2008, pp. 189–214.

Marcus, Alan I., and Howard P. Segal. *Technology in America: A Brief History*. Palgrave, Macmillan Education, 2018.

Marx, Karl. *The Poverty of Philosophy*. Translated by Institute of Marxism Leninism, Progress Publishers, 1955, *Marx/Engels Internet Archive*, www.marxists.org/archive/marx/works/1847/poverty-philosophy/index.htm.

Marx, Leo. "Lewis Mumford: Prophet of Organicism." *Lewis Mumford: Public Intellectual*, edited by Thomas P. Hughes and Agatha C. Hughes, Oxford University Press, 1990, pp. 164–180.

Marx, Leo. "Technology: The Emergence of a Hazardous Concept." *Technology and Culture*, vol. 51, no. 3, 2010, pp. 561–577.

Marx, Leo. *The Machine in the Garden – Technology and the Pastoral Ideal in America*. Oxford Univ. Press, 1964.

Marx, Leo. *The Machine in the Garden: Technology and the Pastoral Ideal in America*. Oxford University Press, 2000.

McLuhan, Eric. "Commonly Asked Questions about McLuhan – the Estate of Marshall McLuhan." Marshall McLuhan, www.marshallmcluhan.com/common-questions/. Accessed 21 February 2019.

McLuhan, Eric, and Marshall McLuhan. *Laws of Media: The New Science*. University of Toronto Press, 1988.

McLuhan, Eric, and Marshall McLuhan. *Media and Formal Cause*. NeoPoiesis Press, 2011.

McLuhan, Eric, and Marshall McLuhan. *Theories of Communication*. Peter Lang, 2011.

McLuhan, Marshall. "Guaranteed Income in the Electronic Age." *Beyond Left & Right: Radical Thought for Our Times*, edited by Richard Kostelanetz, William Morrow & Company, 1968, pp. 72–83.

McLuhan, Marshall. "Introduction" *The Bias of Communication*. University of Toronto Press, 1964.

McLuhan, Marshall. *Letters of Marshall McLuhan*. Edited by Matie Molinaro, Corrine McLuhan, and William Toye, Oxford University Press, 1987.

McLuhan, Marshall. "Printing and Social Change." *Printing Progress: A Mid-Century Report*, The International Association of Printing House Craftsmen Inc., 1959, pp. 81–112.

McLuhan, Marshall. "Technology and Political Change." *International Journal*, vol. 7, no. 3, Summer 1952, pp. 189–195.

McLuhan, Marshall. *The Gutenberg Galaxy – The Making of Typographic Man*. Univ. of Toronto Press, 1962.

McLuhan, Marshall. "The Later Innis." *Queen's Quarterly*, vol. 60, no. 1, Spring 1953, pp. 385–394.

McLuhan, Marshall. "The Mumford Manuscript." 28 February 1973. Fonds, F280, 1990-018, Box 030, File 287. Clara Thomas Archives and Special Collections, Canadian Speakers' and Writers' Service Ltd, York University, Toronto, Ontario, CA. 13 April 2018.

McLuhan, Marshall, et al. *The Gutenberg Galaxy: The Making of Typographic Man*. University of Toronto Press, 2011.

McLuhan, Marshall, and Richard Cavell. *On the Nature of Media: Essays in Understanding Media*. Gingko Press, Inc, 2016.

McLuhan, Marshall, and Terrence W. Gordon. *Understanding Media: The Extensions of Man*. Gingko Press, 2003.
McLuhan, Marshall, and Barrington Nevitt. "The Argument: Causality in the Electric World." *Technology and Culture*, vol. 14, no. 1, 1973, pp. 1–18.
McLuhan, Marshall, and Gerald Emanuel Stearn. "Conversations with McLuhan." *Encounter*, June 1967, pp. 50–57.
Miller, Donald L. *Lewis Mumford: A Life*. Weidenfeld & Nicolson, 1989.
Miller, Donald L. "The Myth of the Machine: I. Technics and Human Development." *Lewis Mumford: Public Intellectual*, edited by Thomas P. Hughes and Agatha C. Hughes, Oxford University Press, 1990, pp. 152–163.
Misa, Thomas J. "How Machines Make History, and How Historians (And Others) Help Them to Do So." *Science, Technology, & Human Values*, vol. 13, no. 3–4, 1988, pp. 308–331.
Mitcham, Carl. *Thinking through Technology: The Path between Engineering and Philosophy*. University of Chicago Press, 1994.
Molella, Arthur P. "The First Generation: Usher, Mumford, and Giedion." *In Context: History and the History of Technology: Essays in Honor of Melvin Kranzberg*, edited by Stephen H. Cutcliffe and Robert C. Post, Lehigh University Press, 1989, pp. 88–105.
Mumford, Lewis. "An Appraisal of Lewis Mumford's 'Technics and Civilization' (1934)." *Daedalus*, vol. 88, no. 3, 1959, pp. 527–536.
Mumford, Lewis. "Apology to Henry Adams." *The Virginia Quarterly Review*, Spring, 1962, pp. 196–217.
Mumford, Lewis. "Authoritarian and Democratic Technics." *Technology and Culture*, vol. 5, no. 1, 1964, pp. 1–8.
Mumford, Lewis. "Civilized History." *New Republic*, vol. 85, no. 1095, November 1935, pp. 63–66.
Mumford, Lewis. *Findings and Keepings*. Harcourt Brace Jovanovich, 1975.
Mumford, Lewis. "History: Neglected Clue to Technological Change." *Technology and Culture*, vol. 2, no. 3, 1961, pp. 230–236.
Mumford, Lewis. *"In Old Friendship" The Correspondence of Lewis Mumford and Henry A. Murray, 1928–1981*. Edited by Frank Novak, Syracuse University Press, 2007.
Mumford, Lewis. *Sketches from Life: The Autobiography of Lewis Mumford: The Early Years*. Dial Press, 1982.
Mumford, Lewis. *Sticks and Stones, A Study of American Architecture and Civilization*. Boni and Liveright, 1924. *Internet Archive*, archive.org/details/sticksstonesstud00mumf/page/n9/mode/2up.
Mumford, Lewis. *Technics and Civilization*. Harcourt, Brace and Company, 1934.
Mumford, Lewis. "The Automation of Knowledge." AV Communication Review, vol. 12, no. 3, 1964, pp. 261–276.
Mumford, Lewis. "The Decline of the West: Form and Actuality (Book)." *New Republic*, vol. 46, no. 597, May 1926, pp. 367–369.
Mumford, Lewis. *The Golden Day*. Horace Liveright., 1926. *Internet Archive*, archive.org/details/in.ernet.dli.2015.215187/page/n5/mode/2up.
Mumford, Lewis. *The Myth of the Machine: Technics and Human Development*. Secker & Warburg, 1967.
Mumford, Lewis. *The Pentagon of Power: The Myth Of The Machine*. Harcourt Brace Jovanovich, 1970.
Mumford, Lewis. *The Story of Utopias*. Boni and Liveright, 1922. *Internet Archive*, archive.org/details/storyutopias00mumfgoog/page/n8/mode/2up.
Mumford, Lewis. "Victor Branford." *New Republic*, vol. 64, no. 821, August 1930, pp. 43–44.
Mumford, Lewis, and Donald L. Miller. *The Lewis Mumford Reader*. Pantheon Books, 1986.
Munslow, Alun. *Narrative and History*. Palgrave Macmillan, 2007.
Munslow, Alun. *The Future of History*. Palgrave Macmillan, 2010.

Noble, David F. *America by Design*. Alfred A. Knopf, New York, 1977.
Novak, Frank, editor. *Lewis Mumford and Patrick Geddes – The Correspondence*. Routledge, 1995.
Nye, David E. *American Technological Sublime*. MIT Press, 1994.
Nye, David E. *Technology Matters: Questions to Live With*. MIT Press, 2006.
"Official Guidebook of the Fair." A Century of Progress, 1933, *Internet Archive*, archive.org/details/officialguideboo00centrich/mode/2up.
Ogburn, William. *Living With Machines*. American Library Association, 1933. *Hathi Trust*, babel.hathitrust.org/cgi/pt?id=mdp.39015065152269&seq=3.
Ogburn, William. *Social Change With Respect to Culture and Original Nature*. George Allen & Unwin Ltd., 1923, *Internet Archive*, archive.org/details/socialchangewit00ogbu/page/n5/mode/2up.
Paul, Herman. "Relations to the Past: A Research Agenda for Historical Theorists." *Rethinking History*, vol. 19, no. 3, July 2015, pp. 450–458.
Paul, Herman, and Anthony Runia. *Key Issues in Historical Theory*. Routledge, 2015.
Polakov, Walter. *The Power Age – Its Quest and Challenge*. Covici and Friede Publishers, 1933.
Post, Robert C. "Back at the Start: History and Technology and Culture." *Technology and Culture*, vol. 51, no. 4, October 2010, pp. 961–994.
Post, Robert C. "Chance and Contingency: Putting Mel Kranzberg in Context." *Technology and Culture*, vol. 50, no. 4, October 2009, pp. 839–872.
Richards, I.A. *Science and Poetry*. Kegan Paul, Trench, Trubner & Co. LTD., 1926. *Internet Archive*, archive.org/details/sciencepoetry0000rich/page/n5/mode/2up.
Richards, I.A., and C.K. Ogden. *The Meaning Of Meaning*. Kegan Paul, Trench, Trubner & Co. LTD., 1930. *Internet Archive*, archive.org/details/in.ernet.dli.2015.218422/page/n5/mode/2up.
Rugg, Harold Ordway. *The Great Technology – Social Chaos and the Public Mind*. John Day Company, 1933. *Hathi Trust*, babel.hathitrust.org/cgi/pt?id=mdp.39015012255702&seq=9.
Schatzberg, Eric. "Technik Comes to America: Changing Meanings of Technology before 1930." *Technology and Culture*, vol. 47, no. 3, 2006, pp. 486–512.
Schatzberg, Eric. *Technology: Critical History of a Concept*. The University of Chicago Press, 2018.
Segal, Howard P. *Future Imperfect: The Mixed Blessings of Technology in America*. University of Massachusetts Press, 1994.
Smith, Allan. "Seven Narratives in North American History: Thinking the Nation in Canada, Quebec, and the United States." *Writing the Nation: A Global Perspective*, edited by Stefan Berger, Palgrave Macmillan, 2007, pp. 63–83.
Smith, Merritt Roe, and Leo Marx, editors. *Does Technology Drive History? The Dilemma of Technological Determinism*. MIT Press, 1994.
Smith, Merritt Roe, and Steven C. Reber. "Contextual Contrasts: Recent Trends in the History of Technology." *In Context: History and the History of Technology: Essays in Honor of Melvin Kranzberg*, edited by Stephen H. Cutcliffe and Robert C. Post, Lehigh University Press, 1989, pp. 133–149.
Sombart, Werner. *Der Moderne Kapitalismus – Die Theorie Der Kapitalistischen Entwicklung*. Zweiter Band. Duncker & Humblot, 1902. *Internet Archive*, archive.org/details/dermodernekapit00sombgoog/page/n6/mode/2up.
Sowell, Thomas. *Marxism: Philosophy and Economics*. Quill, 1985.
Spengler, Oswald. *The Decline of the West*. Translated by Charles Francis Atkinson. George Allen & Unwin Ltd., 1954. *Internet Archive*, archive.org/details/in.ernet.dli.2015.284252/page/n1/mode/2up.
Staudenmaier, John M. "Rationality, Agency, Contingency: Recent Trends in the History of Technology." *Reviews in American History*, vol. 30, no. 1, 2002, pp. 168–181.
Staudenmaier, John M. "SHOT at Fifty." *Technology and Culture*, vol. 50, no. 3, 2009, pp. 623–630.

Staudenmaier, John M. *Technology's Storytellers – Reweaving the Human Fabric*. MIT Press, 1989.
Stoll, Steven. *The Great Delusion: A Mad Inventor, Death in the Tropics, and the Utopian Origins of Economic Growth*. Hill and Wang, 2008.
Susen, Simon. *The Postmodern Turn in the Social Sciences*. Palgrave Macmillan, 2015.
"The Hero." Curb Your Enthusiasm, created by Larry David, performance by Larry David, season 8, episode 6, Home Box Office, 2011.
The Red Deal: Indigenous Action to Save Our Earth. Common Notions, 2021.
Thoreau, Henry D. "Paradise (To Be) Regained." *The Writings of Henry David Thoreau*. Houghton Mifflin Company, 1893, pp. 38–69. *Hathi Trust*, babel.hathitrust.org/cgi/pt?id=uc2.ark:/13960/t6qz23s3c&seq=58.
United States, Department of Health and Human Services, U.S. Surgeon General. *Social Media and Youth Mental Health: The U.S. Surgeon General's Advisory 2023*. Office of the Surgeon General. https://www.hhs.gov/sites/default/files/sg-youth-mental-health-social-media-advisory.pdf.
United States, Senate, Judiciary, Subcommittee on Privacy, Technology, and the Law. *Oversight of A.I.: Rules for Artificial Intelligence*. U.S. Government Publishing Office, 16 May 2023. www.judiciary.senate.gov/committee-activity/hearings/oversight-of-ai-rules-for-artificial-intelligence. Text transcription of hearing.
Veblen, Thorstein. *Absentee Ownership and Business Enterprise in Recent Times – The Case of America*. B.W. Huebsch, 1923. *Internet Archive*, archive.org/details/absenteeownershi0000thor_k2w4/page/n5/mode/2up.
Veblen, Thorstein. *An Inquiry Into The Nature Of Peace And The Terms Of Its Perpetuation*. Macmillan, 1917. *Internet Archive*, archive.org/details/inquiryintonatur00vebl/page/n7/mode/2up.
Veblen, Thorstein. *Imperial Germany and the Industrial Revolution*. Macmillan, 1915. *Internet Archive*, archive.org/details/imperialgermany01veblgoog/page/n6/mode/2up.
Veblen, Thorstein. "On the Nature of Capital I." *The Quarterly Journal of Economics*, vol. 22, 1908, pp. 517–542.
Veblen, Thorstein. "On the Nature of Capital II." *The Quarterly Journal of Economics*, vol. 23, 1908, pp. 104–136.
Veblen, Thorstein. *The Engineers and the Price System*. B.W. Huebsch, 1921. *Internet Archive*, archive.org/details/engineersandpri01veblgoog/page/n7/mode/2up.
Veblen, Thorstein. *The Instinct of Workmanship and the State of the Industrial Arts*. Macmillan, 1914. *Internet Archive*, archive.org/details/instinctworkman00veblgoog/page/n8/mode/2up.
Veblen, Thorstein. *The Place of Science in Modern Civilization and Other Essays*. B.W. Huebsch, 1919. *Internet Archive*, archive.org/details/01610020R.nlm.nih.gov/page/n5/mode/2up.
Veblen, Thorstein. *The Theory of Business Enterprise*. Charles Scribner's Sons, 1904. *Internet Archive*, archive.org/details/theorybusinesse04veblgoog/page/n6/mode/2up.
Veblen, Thorstein. *The Theory of the Leisure Class*. The Modern Library, Random House, 1934. *Internet Archive*, archive.org/details/in.ernet.dli.2015.166218/page/n7/mode/2up.
Veblen, Thorstein. *The Vested Interests and the State of the Industrial Arts*. B.W. Huebsch, 1919. *Internet Archive*, archive.org/details/vestedinterestss00vebliala/page/n5/mode/2up.
Watson, Rita, and Menahem Blondheim, editors. *The Toronto School of Communication Theory: Interpretations, Extensions, Applications*. University of Toronto Press, 2007.
Weizenbaum, Joseph. *Computer Power and Human Reason: From Judgement to Calculation*. W.H. Freeman & Co Ltd., 1976.
Wells, H. G. "Civilization at the Breaking Point." *New York Times*, 27 May 1915, p. 10.
Westbrook, Robert. "Tribune of the Technostructure: The Popular Economics of Stuart Chase." *American Quarterly*, vol. 32, no. 4, 1980, pp. 387–408.

White, Hayden. “Historiography and Historiophoty.” *The American Historical Review*, vol. 93, no. 5, December 1988, pp. 1193–1199.

White, Hayden. *Metahistory: The Historical Imagination in Nineteenth-Century Europe*. Johns Hopkins University Press, 1975.

White, Hayden. “The Future of Utopia in History.” *Historein*, vol. 7, May 2008, pp. 12–19.

White, Hayden. *The Practical Past*. Northwestern University Press, 2014.

Williams, Rosalind. “Classics Revisited: Lewis Mumford’s *Technics and Civilization*.” *Technology and Culture*, vol. 43, no. 1, 2002, pp. 139–149.

Williams, Rosalind. “Lewis Mumford as a Historian of Technology in *Technics and Civilization*.” *Lewis Mumford: Public Intellectual*, edited by Thomas P. Hughes and Agatha C. Hughes, Oxford University Press, 1990, pp. 43–65.

Winant, Gabriel. “We Live in a Society.” *N+1*, www.nplusonemag.com/online-only/online-only/we-live-in-a-society. Accessed 06 June 2023.

Wyatt, Sally. “Technological Determinism Is Dead; Long Live Technological Determinism.” *The Handbook of Science and Technology Studies*, edited by Edward J. Hackett, et al., MIT Press, 2008, pp. 165–180.

Zuboff, Shoshana. *The Age of Surveillance Capitalism: The Fight for a Human Future at the New Frontier of Power*. Public Affairs, 2019.

Appendix 1
The U.S. Origins of the Concept Technology

It was Eric Schatzberg who outlined the specific U.S. origins of the concept.[229] He and others have demonstrated that the term technology is a rather ambiguous concept.[230] It has a broader meaning than machine, and that meaning has changed or shifted a few times since it first few invocations. By the nineteenth century, the Greek-derived term had come to be used in English and other European languages to refer to the "description, principles, or teaching of the practical arts and, less often, to technical terminology" (Schatzberg, "Technik" 489). It was used to denote a field of study. According to Schatzberg's research, it is not Jacob Bigelow's 1829 book *Elements of Technology*, but the later establishment of "the Massachusetts Institute of Technology in 1861 that brought technology into more common usage in the United States" ("Technik" 491–492). And although a more standard term (polytechnic) was available at the time of its founding, thereafter the word technology started to win out as a way to describe a school of higher technical education.

The change from an obscure term that primarily was associated with education to a more all-encompassing description of an object of industry took place even more recently. Schatzberg's argument has a broad and a more proximate aspect. First, in agreeing with Leo Marx, he cites "the 'semantic void' created by the growth of large-scale technological systems" ("Technik" 493). This more general background holds that other terms were not adequate to describe the changing landscape of everyday experience. More precisely, Schatzberg then homes in on a "semantic shift [that] took form during the early decades of the twentieth century." His findings reveal that the new connotations were mainly taken from the writings of American social scientists. As previously noted, these authors imbued and conflated the English word technology with the German discourse of the word *Technik*. The result was a semantic turn from its original definition as "the science or study of the useful arts" and its associated meaning of a school of technical education, to a new meaning that "includes the industrial arts as a whole, including the material means of production" (487).

Parallel to these semantic shifts is the underlying and ever-increasing presence of machines in daily life. The power to shape and mold environments could now be attributed to technology, thus increasing the amount of semantic clout

229 For a more complete, but compact overview of the history of the concept see chapter one of Eric Schatzberg's *Technology – Critical History of a Concept*, pp. 8–13.
230 On the ambiguity of the term technology see also Leo Marx, "Technology," p. 574, and David E. Nye, *Technology Matters*, p. 15.

https://doi.org/10.1515/9783111396477-010

the term was, and still is, able to command. Included in contemporary definitions of technology is both the potential and real capability of knowledge/information, its application, a manner of accomplishing something using technical processes and the final material outcome, frequently an object.

In the current book, technology should mainly be regarded in its functional, if "hazardous," role as a historical subject.[231] The hazards, as Leo Marx has described, involve the concept of technology's "lack of specificity," its tendency to be "represented by things," and its status in many cases as private property ("Technology" 576). These factors should not distract from the fact that invocation of the concept should fundamentally be understood to involve human agency and socio-economic and cultural relations between people.

231 Leo Marx elaborates on why the concept of technology is hazardous at the end of his essay "Technology – Emergence of a Hazardous Concept."

Appendix 2
Four Different Common Uses of the Word History

This appendix details four different common understandings of the word history: history as the subject matter of historians, history as account, historical reality, and history as the past. Such an exercise in semantics is important in order to counter the tendency in common language usage to conflate the different meanings of the word. These four meanings have bearing on interpretations of the technology-drives-history narrative.

History in the first sense involves a historian's practice in which they theoretically abstract from historical realities. For the purposes of the current study, it should be understood as the discipline that produces historiography, or the written representation of the past, and of what Hayden White calls "historiophoty" ("Historiography and Historiophoty"), namely, the depiction of the past in images in videos, films, and photos. A common attribute that both historiography and historiophoty share is that they are mainly performed by professional historians, local guides, and other interested parties.[232] This categorization obviously includes the author-historians analyzed in this study. Veblen, Mumford, Chase and McLuhan should therefore be viewed not as "committed to the reality of the past [. . .] and independent of all utilitarian demands," but rather as inventors of collocated, subjective written accounts that seek to represent and then justify a meaningful narrative of a technological past (Munslow, *The Future of History* 33–35). Consideration of their voices alerts readers to the ontological independence of historical reality from the historiography they have constructed. The simple fact that their histories are printed books clearly demonstrates the non-trivial difference between the second meaning of the word history – history as account – and other concepts such as the past and historical reality.

Not even the most "global sceptic" would deny historical reality (Munslow, *The Future of History* 40). People live, objects exist, things happen. But material and social realities go out of existence, change meaning, or only leave their traces. The resulting "epistemological problem" arises when a "conventional historian" of technology follows a strict methodology that attempts to find correspondence between archived sources or material traces and an empirical truth – historical reality – that should emerge from the historical evidence (15–32). According to Munslow, it is only when the "irony of historical authorship" is accepted that readers of technology-history can understand the large array of "authorial arti-

232 A historian in a general sense is therefore anyone who is sanctioned to perform or author an account of the past.

https://doi.org/10.1515/9783111396477-011

fice"[233] that actually creates the meaning of a narrative as opposed to a mimesis of historical reality (276). What constitutes a historical account in the history of technology is often by definition, value-laden with the ideas brought to the subject by author-historians and the choices that they make.

In contrast to historical reality, the fourth meaning invoked by the word history – the past – is a mental build of time, a "story space" that can accommodate the author-historian's "fictively constructed narrative model" of what happened or of how things were (Munslow, *Narrative and History* 144). Theorist Herman Paul mentions at least four types of past that can be distinguished based on various properties. The first type can be stretched to deal with times past, present and future and ordered sequentially as in the clock and calendar-defined "chronological past" (Paul and Runia 19–20). Clearly distinguishing between the three times can prove tricky, so author-historians often hold to conventions like the directionality and uniformity of time to clearly define what is from what was and what will be (20). The idea of progress in history as understood in modern historiography clearly borrows from this notion of the past. It allowed many engineering minds to calculate the positive steps toward a better time by implicitly drawing the line between a monarchial-European past and a technical-utopian American future.

A second type of past is one that is layered or divided into epochs to signal its "development and individuality" (Lorenz and Bevernage 25). An example of this type of past is when author-historian McLuhan unknowingly invokes the so-called "completed past"[234] in announcing to contemporary readers that they had entered the electronic age.[235] Through periodization or the "breaking up time" into such epochs, an "active producing" of a model of the past, meaning the role an author-historian assumes in constructing a narrative about historical reality, can be readily observed.

A third type of past is characterized by its otherness compared to the present (Paul and Runia 23–25). This "strange past," as it has been termed, alludes to the feeling of discrepancy between present and past conditions. One could imagine

---

233 Trickery that includes "voice, focalization, timing, characterization, emplotment" etc. (Munslow, *The Future of History* 276).

234 Paul associates the ideas of a completed past with the nineteenth-century idea of Historicism, such as thinking about the process of historical development toward individuality. According to this concept, the content of the completed past does not belong to the present (Paul and Runia 21–23).

235 See the first and second introduction to *Understand Media*, where McLuhan orders the characteristic "implosion" of the electronic age after 3000 years of "explosion" of the mechanical age (McLuhan and Gordon 5–16).

that, sensing an increase in the effects of technology, conclusions are drawn that the past is radically different from current conditions. It is then part of an author-historian's relation to a technological past that they attempt to show continuity over time, demonstrate the development of technologies and thus reduce the anxiety and assuage feelings of alienation. An author that tackles this feeling directly in his seminal book *Machine in the Garden* is Leo Marx, whose literary trope by the same name stands for the disturbance of the pastoral ideal by the introduction of various technologies. A concrete example of this change in perception is the image of a locomotive or train speeding through a beautiful rural landscape. The dialectical negotiation of these two symbols in nineteenth-century American literature forms the content of the book.

A fourth type of past, the "present past," creeps into the present when an account in a book, film, architecture, or memory is read, presented, looked at, or recalled ensuring that the past "lives on unconsciously in the present" (Paul and Runia 25–27). The continuity from past to present can be observed through the reification of a historical subject. In the specific case of using technology as the subject of an account of the past, it is therefore crucial to understand how technology as a concept has been construed by an author-historian and how a meaning of technology shows continuity from past to present. Mumford's famous example of the "big machine" or "megamachine," formerly composed of human parts, shifting towards "more reliable mechanical parts" as man became increasingly mechanized, is an invention that he elevates to the role of subject to describe human history over millennia (*The Myth of the Machine* 188). As such, the past is made present through a historical subject that links the two.

The four different faces of history that have just been presented illustrate the various ways in which the question – does technology drive history? – can be interpreted. Differentiating the types helps to illustrate how author-historians construct an account, not of historical reality, but of a mental model of time in a "story space" commonly referred to as "the past."[236] Through their performed historiographical accounts and individual narrative choices, the authors examined in this book all construct different models of a preferred past.

236 According to Munslow, the past-as-history refers to the relationship of the past (content) through the process of its construction by an author-historian (narration) in an expression or account (form). Ultimately, as he points out, understanding this relationship between the three elements above leads to a definition of history as a story space (*Narrative and History* 26–27). For example, if the form is changed, in other words history is substituted for, say, chronicles or annals, the past becomes very different (White, *Metahistory* 5–7, 60–62).

Index

https://doi.org/10.1515/9783111396477-012